GLOBAL CAPITALISM AND CLIMATE CHANGE

Globalization and the Environment Series

Series Editors

Richard Wilk, Department of Anthropology, 130 Student Building, Indiana University, Bloomington, IN 47405, USA, or wilkr@indiana.edu

Josiah Heyman, Department of Sociology & Anthropology, Old Main Building #109, University of Texas, 500 West University Avenue, El Paso, TX 79968, USA, or jmheyman@utep.edu

This AltaMira series publishes new books about the global spread of environmental problems. Key themes addressed are the effects of cultural and economic globalization on the environment; the global institutions that regulate and change human relations with the environment; and the global nature of environmental governance, movements, and activism. The series will include detailed case studies, innovative multi-sited research, and theoretical questioning of the concepts of globalization and the environment. At the center of the series is an exploration of the multiple linkages that connect people, problems, and solutions at scales beyond the local and regional. The editors welcome works that cross boundaries of disciplines, methods, and locales and span scholarly and practical approaches.

Books in the Series

Power of the Machine: Global Inequalities of Economy, Technology, and Environment, by Alf Hornborg (2001)

Confronting Environments: Local Environmental Understanding in a Globalizing World, edited by James Carrier (2004)

Communities and Conservation: Histories and Politics of Community-Based Natural Resource Management, edited by J. Peter Brosius, Anna Lowenhaupt Tsing, and Charles Zerner (2005)

Globalization, Health, and the Environment: An Integrated Perspective, edited by Greg Guest (2005)

Cows, Kin, and Globalization: An Ethnography of Sustainability, by Susan A. Crate (2006)

Global Visions, Local Landscapes: A Political Ecology of Conservation, Conflict, and Control in Northern Madagascar, by Lisa L. Gezon (2006)

Globalization and the World Ocean, by Peter Jacques (2006)

Rethinking Environmental History: World-System History and Global Environmental Change, edited by Alf Hornborg, John McNeill, and Joan Martínez-Alier (2007)

The World's Scavengers: Salvaging for Sustainable Consumption and Production, by Martin Medina (2007)

Saving Forests, Protecting People? by John W. Schelhas and Max J. Pfeffer (2008)

Capitalizing on Catastrophe: Neoliberal Strategies in Disaster Reconstruction, edited by Nandini Gunewardena and Mark Schuller (2008)

World in Motion: The Globalization and the Environment Reader, edited by Gary M. Kroll and Richard H. Robbins (2009)

War and Nature: The Environmental Consequences of War in a Globalized World, by Jurgen Brauer (2009)

Computing Our Way to Paradise? The Role of Internet and Communication Technologies in Sustainable Consumption and Globalization, by Robert Rattle (2010)

Global Capitalism and Climate Change: The Need for an Alternative World System, by Hans A. Baer (2012)

GLOBAL CAPITALISM AND CLIMATE CHANGE

The Need for an Alternative World System

Hans A. Baer

ALTAMIRA PRESS

A Division of Rowman & Littlefield Publishers, Inc.
Lanham • New York • Toronto • Plymouth, UK

Published by AltaMira Press
A division of Rowman & Littlefield Publishers, Inc.
A wholly owned subsidiary of The Rowman & Littlefield Publishing Group, Inc.
4501 Forbes Boulevard, Suite 200, Lanham, Maryland 20706
www.rowman.com

10 Thornbury Road, Plymouth PL6 7PP, United Kingdom

British Library Cataloguing in Publication Information Available

Library of Congress Cataloging-in-Publication Data

Baer, Hans A., 1944-
Global capitalism and climate change : the need for an alternative world system / Hans A. Baer.
p. cm. — (Globalization and the environment series)
Includes bibliographical references and index.
ISBN 978-0-7591-2132-4 (cloth : alk. paper) — ISBN 978-0-7591-2134-8 (ebook)
1. Climatic changes—Economic aspects. 2. Globalization. 3. Sustainable development. I. Title.
QC903.B14 2012
330.12′6—dc23

2012016730

™ The paper used in this publication meets the minimum requirements of American National Standard for Information Sciences—Permanence of Paper for Printed Library Materials, ANSI/NISO Z39.48-1992.

Printed in the United States of America

Contents

Preface

Over the course of the past several years, a growing number of anthropologists, as well as other social scientists, including sociologist Anthony Giddens (2009), have turned their attention to climate change or global warming. Roncoli and Magistro (2000) had urged anthropologists to examine global climate change as part and parcel of the anthropology of climate variability, a phenomenon that includes droughts, hurricanes, and other instances of erratic weather patterns. While archaeologist Brian Fagan (1999:76) is correct in his seemingly dismissive assertion that "global warming is nothing new for humanity," the magnitude of warming that the planet has been experiencing, particularly in the past several decades, and that the vast majority of climate scientists predict will occur throughout the present century and beyond (even if it could be checked by monumental preemptive measures) is on a magnitude never experienced by humanity, in part due to the fact that there never have been so many people inhabiting so many places in our fragile biosphere. He has been discussing the impact of climate change, albeit of a more natural form than an anthropogenic one, on human societies for some time.

Of investigations into climate change in more recent times, a notable effort is an anthology titled *Anthropology and Climate Change: From Encounters to Actions*, edited by Susan A. Crate and

Mark Nuttall (2009). This book is a welcome addition to the still-emerging anthropology of climate change. Yet, a major shortcoming of this book, and of most of the anthropological work on climate change thus far, is that it fails to view climate change as yet another contradiction of global capitalism with its treadmill of production and consumption heavily reliant on fossil fuels and its commitment to ongoing economic expansion, regardless of the social and environmental consequences. In *Global Warming and the Political Ecology of Health*, published shortly before Crate and Nuttall's anthology, Merrill Singer and I adopted a critical anthropological perspective in examining the impact of climate change on health. This present book seeks to go beyond that earlier one in delineating the roots of climate change in global capitalism and the systemic changes needed to create a more socially just and environmentally sustainable world system that would move humanity toward a safer climate. In this effort, my approach is more that of a historical social scientist who happens to have a PhD in anthropology than of an anthropologist in the conventional sense of the word. In this effort, I have been guided by the work of an array of political ecologists and eco-Marxists, particularly John Bellamy Foster (2000, 2009) (an environmental sociologist trained in political economy at the University of Oregon and the editor of *Monthly Review*), Joel Kovel (2007), Ariel Salleh (2009), and contributors to the journal *Capitalism Nature Socialism*.

Acknowledgments

My scholarly interest in climate change or global warming began in the hot summer of 2005 while working on the first edition of *Introducing Medical Anthropology* (AltaMira Press, 2007) with Merrill Singer. In chapter 7 of our textbook on "Health and the Environment" we included a section on "The Impact of Global Warming on Health." Indirectly this small effort led to a book titled *Global Warming and the Political Ecology of Health* (Left Coast Press, 2009), the sixth book that we had done together. Merrill and I became acquainted as graduate students in late 1975 in the anthropology department at the University of Utah and we have remained close friends, colleagues, and comrades in the struggle for social justice and environmental sustainability ever since, despite the geographical distance that separates us with him residing in Storrs, Connecticut, and me in Melbourne, Australia. Since coming to Melbourne, I have become a friend and colleague of Verity Burgmann in the School of Social and Political Sciences. We have written a book titled *Australian Climate Politics and Climate Movement* (Melbourne University Press, 2012). I owe much to Verity as an immigrant and soon-to-become Australian citizen in acquainting me with Australian politics and social movements. Both of us are partisan observers of the Australian climate movement.

Upon arriving at the University of Melbourne in January 2006, I quickly touched base with various fellow academics as well as students who share an interest in climate change. They include Jon Barnett, Peter Christoff, Liam Cooper, Peter Dwyer, Robyn Eckersley, Peter Ferguson, Jim Falk, Melanie Lowe, Anthony Marcus, Monica Minnegal, Thomas Reuter, and Alan Thorold. While visiting Melbourne as a research fellow in 2007, Kay Milton, a renowned environmental anthropologist who now resides in New Zealand, participated in a symposium on "The Impact of Global Warming on the Environment and Human Societies" that I convened on April 20, 2007, at the University of Melbourne. Other presenters at that symposium included Peter Christoff, Jim Falk, Janet McCalman, Murry Peel, A. Barrie Pittock, and Murray Peel. Fellow co-convenors of panels on climate change at annual conferences of the Australian Anthropological Society have included Marcus Barber, Megan Jennaway, Kay Milton, and Thomas Reuter.

I would also like to acknowledge a number of researchers and climate activists who have shaped my understanding of climate change and climate politics. They include Ian Angus, Fiona Armstrong, Sue Bolton, Simon Butler, SallyRose Carbines, Ben Courtice, Chris Breen, John Bellamy Foster, Jeremy Moss, Judy McVey, Andrew Milner, Jane Morton, Dick Nichols, Bronwyn Plarre, Thomas Reuter, John Rice, Ariel Salleh, David Spratt, Philip Sutton, Ted Trainer, Cam Walker, and Erik Olin Wright. I would also like to acknowledge many members of the Socialist Alliance, Climate Action Moreland, the Climate Emergency Action Network who attended workshops and presentations that I did on climate change-related topics and Dominique Finney who facilitated my doing a climate change-related talk at the Woodford Folk Festival in Queensland in December 2009. I extend appreciation to the University of Melbourne for granting me a six-month study leave in 2009 to conduct research on Australian climate politics and the climate movement. I also acknowledge the contribution of Wendi Schnaufer and Elaine McGarraugh, my editors at AltaMira who so patiently assisted in bringing this book to completion and the input of various anonymous peer reviewers in shaping it. Last but not least I would like to talk my children, Eric and Andrea, for listening to their dad either in person or by email go on about the critical anthropology of climate change.

Introduction

Numerous climate scientists have come to the conclusion that climate change is largely the result of human or anthropogenic activities, particularly since the Industrial Revolution. In short, climate change has already had severe economic, political, and health consequences for humanity and will continue to do so as the twenty-first century unfolds. Human societies have never faced an environmental problem on this scale before. Climate change and its repercussions have become topics of increasing public awareness, although this awareness varies considerably from society to society as well as within societies. For example, the discourse on climate change tends to be much more marked in Europe than it is in the United States and Australia for that matter, two countries where I have resided at length, the first for about 50 years and the latter for about 7 years. Awareness of abrupt climate change has found its way into popular culture, the mainstream media, and science fiction. Al Gore's movie *An Inconvenient Truth* and accompanying book (Gore 2006) and the *Stern Report* authored by Nicholas Stern (2007), a former World Bank economist, in particular propelled climate change into public consciousness around the world. A growing number of business leaders and politicians have come to embrace a form of *green capitalism,* which asserts that climate change poses a serious threat to the existing global economy but that capitalism has the capacity to reform itself, adopt new forms of energy and environmentally sustainable technologies, and continue to sustain economic expansion and profit making. Conversely, various radical environmentalists, eco-socialists, and certain critical social scientists view climate change as yet one more manifestation of the contradictions, perhaps the most profound contradiction, of global capitalism.

While humans indeed have been emitting greenhouse gases for some time, the Industrial Revolution with its heavy reliance on fossil fuels and the capitalist treadmill of production and consumption contributed to a new type of climatic change, one generated not so much by natural events as by human-induced, or anthropogenic, activities, as numerous climate scientists have concluded. Brian Fagan (2008:xvii) asserts that "we've entered a time of sustained warming, which dates back to at least 1860, propelled in large part by humanity—by the greenhouse gases from fossil fuels." Elsewhere, William Ruddiman (2005:171) offers a caveat to this contention by noting that "beginning in the late 1800s, use of fossil fuels (first coal, and later oil and natural gas) rapidly increased, eventually replacing deforestation as the primary source of CO_2 emissions by humans."

While climate scientists debated for a long time whether recent climate change has been primarily a natural phenomenon rather than an anthropogenic one, the vast majority of them now agree that it has been largely created by the emission of various greenhouse gases, particularly carbon dioxide (CO_2), which has increased from 280 parts per million (ppm) at the time of the Industrial Revolution to 390 ppm in 2010. In contrast, the level of CO_2 "varied between a minimum of 180 ppm and a maximum of 280 ppm," with the lower levels having occurred during glacial periods and the higher levels during interglacial periods over the course of some 400,000 years before 1800 (Ward 2010:56). As Renee Hetherington and Robert Reid (2010:269) astutely observe, "Our growing obsession with, and economic dependency on, fossil fuels, combined with our penchant for consumerism, has resulted in humans becoming a climate-change mechanism."

In short, anthropogenic climate change has been inducing, and will continue to induce, severe economic, social, political, military, and health consequences as the twenty-first century unfolds. The Australian Academy of Sciences (2010:3) reports that climate models "estimate that by 2100, the average global temperature will be between 2°C and 7°C higher than pre-industrial temperatures, depending on future greenhouse gas emissions and on the ways that models represent the sensitivity of climate to small disturbances." While most projections of climate change tend to focus on the twenty-first century, climate models also indicate that climate change will continue well after 2100. Given that humanity has been on the face of the planet for some 5 to 6 million years, ongoing global warming and associated climatic changes raise questions about how long humanity can thrive—at least in its present numbers and occupying as much of the Earth as it does today—into and beyond the twenty-second century. As the Australian Academy of Sciences (2010:3) so aptly observes,

> A warming of 7°C would greatly transform the world from the one we now inhabit. . . . Such a large and rapid change in climate would likely be beyond the adaptive capacity of many societies and species.

Some scholars refer to the period in which greenhouse gas emissions began to build up as the Anthropocene. Ruddiman (2005:5) contends that CO_2 emissions began to slowly increase as humans began to clear the land in their shift from foraging to farming about 8,000 years ago in places such as China, India, and Europe. Starting about this time, the burning of peat for heating and cooking and of limestone to produce lime for mortar and plaster also added to CO_2 emissions. Ruddiman contends that methane (CH_4) emissions began to increase around 5,000 years ago as various populations started to irrigate for rice production and raise livestock. Livestock produces methane both from manure and gaseous belches. The clearing of forests and burning of grasslands also produced methane as did human waste. Ruddiman (2005:64) asserts that greenhouse gases emitted by anthropogenic activities have created a "warming effect that counteracted most of the natural cooling" and in essence "stopped a small-scale glaciation that would have naturally developed in far northeastern Canada."

Various progressive scholars, particularly in the social sciences, have increasingly come to acknowledge that anthropogenic climate change, or global warming, that has been occurring at least since the Industrial Revolution constitutes yet another contradiction of global capitalism. In an increasingly globalized economic system, the capitalist drive for profit making and economic expansion results in a perpetual treadmill of production and consumption heavily reliant on fossil fuels and other substances that produce greenhouse gas emissions. While John Bellamy Foster acknowledges that climate change constitutes the most serious ecological threat impacting upon both humanity and the planet, he views it as a manifestation of a larger global environmental crisis with its interrelated components. Foster (2010:3) asserts, "Independently of climate change, tropical forests are being cleared as a direct result of the search for profits. Soil destruction is occurring due to current agribusiness practices. Toxic wastes are being diffused through the environment. Nitrogen run-off from the overuse of fertilizer is affecting lakes, rivers, and ocean regions, contributing to oxygen-poor dead zones."

While physical scientists have tended to dominate the discourse on climate change, it is imperative that social scientists, especially critical ones, engage in scholarly activity on the most crucial environmental issue of our time. As Peter Grimes and Jeffrey Kentor (2003:261) argue, physical scientists generally "cannot address the political, economic, and social forces that explain the choice of systems, machinery, and locations employing compounds responsible for global warming." Bearing this thought in mind, it is imperative that social scientists, including anthropologists, give greater consideration to climate change than has tended to be the case thus far. While I am primarily an anthropologist, I recognize that the effort to examine the impact of climate change or global warming on humanity has to be an interdisciplinary effort, one that involves collaboration among climate and other natural scientists, social scientists, public health people, policy analysts, and humanists.

From my base as a transplanted American in Melbourne, Australia, I have been engaged since early 2006 in the development of a critical anthropology of global warming or climate change (Baer 2007, 2008, 2009; Baer and Singer 2009). Hopefully, anthropologists and other social scientists, along with progressive climate scientists, can contribute to a larger effort not only to mitigate the impact of climate change on humanity but also to envision and struggle for an alternative world system, one committed to meeting people's basic needs and striving for social equity, justice, and environmental sustainability. Like the social sciences, as Steven Vanderheiden (2008) observes, climatology has the potential to serve as a form of social critique instead of acting as a largely descriptive effort. We have seen that various climate scientists, such as James Hansen in the United States and David Karoly in Australia, have become vocal climate activists. At the same time, climate science thus far, as a form of social critique, has been very limited, as is exemplified by the fact that the mitigation strategies of Working Group 3 of the Intergovernmental Panel on Climate Change (IPCC) have been framed within the parameters of global capitalism. Furthermore, corporations and politicians, while acknowledging the reality of anthropogenic climate change, often ultimately downplay or ignore climate science scenarios for the future and continue with "business as usual." As Tim Luke (2008:146) observes,

> Good science with reliable finding about global warming trends has been available for decades. Yet, during these same decades, very little has been done effectively to reduce net greenhouse gas emissions beyond identifying and aiming at a future ceiling level pegged to floor values measured in 1990.

Global Capitalism and Climate Change constitutes an effort to develop a critical social science of climate change, one that posits its roots in global capitalism with its treadmill of production and consumption, heavy reliance on fossil fuels, and commitment to ongoing economic expansion. Furthermore, this book explores the systemic changes necessary to create a more socially just and sustainable world system that would possibly start to move humanity toward a safer climate, as well as the role of a burgeoning climate movement in this effort.

Anthropogenic climate change has been inducing and will continue to induce severe economic, social, political, military, and health consequences as the twenty-first century unfolds. Anthropologists have often noted that social systems, whether at the local, regional, or global level, do not last for effort. Thus, perhaps more than any other environmental crisis, anthropogenic climate change forces us to examine whether humanity needs to transcend global capitalism and develop an alternative, or, more precisely, a democratic eco-socialist world system.

Chapter 1 provides an overview of the impact on the environment of climate change induced by various greenhouse gases, particularly carbon dioxide, methane, and nitrous oxide (N_2O). It summarizes the findings of climate science with respect to the impact of climate changes on (1) rising temperatures; (2) rising sea levels, warming oceans, and melting ice caps, glaciers, and tundras; (3) erratic weather patterns; (4) biodiversity; and (5) safe temperatures and tipping points. This chapter also summarizes the impact of climate change on human societies, particularly settlement patterns, subsistence and food security, and health. Once various feedback chains, related in part to the long lifetime of some greenhouse gases, get started, they may be self-perpetuating and need no further anthropogenic input to keep going.

In chapter 2, I argue that climate change constitutes one of the most important issues—perhaps the most important issue in that it is related to numerous other issues—of the twenty-first century. This chapter explores the following contradictions of the capitalist world system: (1) its emphasis on profit making, economic expansion, and the treadmill of production and consumption; (2) the growing socioeconomic gap between rich and poor both within and between nation-states; (3) the depletion of natural resources and environmental degradation, the most profound form of which is climate change; (4) population growth, which in large part is stimulated by ongoing poverty; and (5) the resource wars of various developed countries, particularly the United States, the United Kingdom, and Australia, in doing the bidding of multinational corporations.

Chapter 3 focuses on the capitalist treadmill of production and consumption as a source of greenhouse gas emissions, which in turn contribute to anthropogenic climate change. This, perhaps more than any other environmental crisis, illustrates the unsustainability of the capitalist world system. Various world systems theorists have examined the linkage between a nation-state's position in the capitalist world system and its environmental impact, including on climate change. While energy efficiency has tended to improve in core countries, there has also been a tendency for total carbon dioxide emissions and per capita emissions to increase. Such a trend is consistent with the Jevons paradox, which observes that despite technological improvements under capitalism, with its emphasis on economic expansion, there is a tendency toward increasing energy consumption. This chapter discusses in detail the following sources of greenhouse gas emissions within the context of global capitalism: (1) fossil fuels, namely, coal, petroleum, and natural gas; (2) steel, aluminum, and cement/concrete production; (3) transport, particularly motor vehicles, airplanes, and marine shipping; (4) housing units and buildings; (5) a seemingly endless array of consumer items; (6) industrial agriculture and logging; and (7) militarism and wars. It also examines the ecological footprints and greenhouse gas emissions of the "big two"—the United States and China—as well as the United Arab Emirates.

In chapter 4, I explore the inadequacies of existing climate regimes as purported climate change mitigation strategies. While it is inevitable that over the short run humanity will have to adapt to climate change, the more crucial issue is that of mitigation—that is, transcending climate change in order to ensure the survival of humanity as well as preserve biodiversity. Since the late 1980s, climate regimes have emerged at the international, regional, provincial, state, and even local levels. The vast majority of climate regimes function within the parameters of green capitalism—the notion that capitalism, by adopting emissions trading schemes, various technological innovations, energy efficiency, recycling, and other practices, can be environmentally sustainable. This chapter highlights the limitations of existing climate regimes, such as the Kyoto Protocol and the EU Emissions Trading Scheme.

Chapter 5 focuses on the limitations of green capitalism or climate capitalism in mitigating climate change. While historically corporations have been resistant to the assertion on the part of environmental activists that many of their practices are environmentally destructive and also contribute to climate change, a growing number of corporations have begun to acknowledge that they can make advances in sustainable development while reducing their greenhouse gas emissions by engaging in a process of ecological modernization. While technological innovations, such as renewable sources of energy and energy efficiency, have an important role to play in climate change mitigation, even they cannot contain climate change over the long run as long as they accept the capitalist imperative for continual economic growth.

In chapter 6, I propose the vision of a global democratic eco-socialist system as an alternative to the existing capitalist world system. Climate change compels us to engage in a serious assessment of alternatives to global capitalism. Before engaging in such an exercise, I discuss various dystopian visions of the future discussed by Mark Lynas in his book *Six Degrees*, by James Lovelock in various books, and by proponents of eco-authoritarian regimes. This chapter also explores various social justice initiatives that, while not seeking to transcend global capitalism per se, seek to make it both more socially just and environmentally sustainable, including in terms of climate change. Conversely, I maintain that it is imperative to think outside the box and construct an alternative to global capitalism as the ultimate climate change mitigation strategy. Thus, I propose the creation of a democratic eco-socialist world system as a form of what Erik Olin Wright terms a *real utopia*. Despite efforts in the Soviet Union, China, and numerous other postrevolutionary societies to create socialism, all attempts to achieve this ideal were hindered by complex historical and social structural conditions. Democratic eco-socialism remains a vision that would entail the following dimensions: (1) an economy oriented toward meeting basic social needs, including adequate food, clothing, shelter, and health; (2) a high degree of social equality; (3) public ownership of productive forces; (4) representative and participatory democracy; and (5) environmental sustainability. Indeed, developments in Latin America raise the hope of creating "socialism for the twenty-first century." Ultimately, the shift to democratic eco-socialism in any country would have to be part of a global process that no one can fully envision at this time.

Chapter 7 explores various transitional progressive reforms that potentially would pave the way to an alternative world system committed to social justice and environmental sustainability. Obviously, the transition toward a democratic eco-socialist world system is not guaranteed and will require a tedious, even convoluted path. Nevertheless, while awaiting the revolution, so to speak, progressive people can work on various transitional reforms. In this chapter, I propose the following transitional reforms essential to implementing an ecological revolution and ultimately global democratic eco-socialism: (1) the creation of new left parties; (2) the implementation of emissions taxes at sites of production that include efforts to protect low-income people; (3) the socialization in various ways of the means of production; (4) increasing social equality within nation-states and between nation-states; (5) the implementation of workers' democracy; (6) the shortening of the workweek; (7) the adoption of renewable energy sources, energy efficiency, and appropriate technology and the creation of green jobs; (8) the expansion of public transport; (9) the creation of green cities; (10) resistance to the capitalist culture of consumption; and (11) the creation of sustainable agriculture and forestry. The transitional steps that I have delineated constitute a loose blueprint for shifting human societies or countries toward democratic eco-socialism and a safe climate, but it is important to note that both of these phenomena will entail a global effort, including the creation of a progressive climate governance regime.

In chapter 8, I examine the emerging climate movement, which I view as a disparate but potentially antisystemic development. The climate movement, both internationally and nationally, is a broad phenomenon that draws in part upon earlier movements, particularly the environmental movement but also the global justice or anti–corporate globalization, indigenous rights, and labor movements. It encompasses the following tendencies: (1) a green social democratic tendency that emphasizes ecological modernization; (2) a radical, anticapitalist tendency that seeks drastic systemic change; and (3) an in-between tendency that recognizes social justice issues but is not explicitly anticapitalist. While touching upon the US climate movement, this chapter, based in large part on my own ethnographic research, discusses how these tendencies play themselves out in the Australian climate movement. This chapter ends with a call for transforming the climate movement both internationally and in its various national manifestations into a climate justice movement that calls for the transcendence of global capitalism and a shift to an alternative world system based on social parity and environmental sustainability. It discusses efforts in making such a shift as manifested in the Durban Group for Climate Justice, Climate Justice Now!, Klimaforum at the Copenhagen Climate Conference in December 2009, and the World People's Conference on Climate Change in April 2010 in Bolivia.

In my concluding chapter I argue that the effort to examine and mitigate the impact of climate change on humanity must be an interdisciplinary one that involves collaboration among natural and social scientists, public health people, policy analysts, and humanists who are willing to collaborate with the climate justice movement and other antisystemic movements. Going from the present capitalist world system—which has generated, and continues to generate, anthropogenic climate change—to an alternative global political economy, however it is defined, will require much effort. And there are no guarantees that we will be able to create a more socially equitable and environmentally sustainable world. But do we really have any other meaningful choice than continuing on a downward spiral that threatens the destruction of much of humanity and other forms of life as well as further environmental degradation, including climate change?

1

The Impact of Climate Change on the Environment and Human Societies

When one contemplates time in terms of the age of the universe (estimated to be around 15 billion years old) or even our planet (estimated to be about 5 billion years old), one quickly realizes that our existence as a species has been so far, and probably will ultimately constitute, a quick blip. Gareth Morgan and John McCrystal (2009:85–86) delineate a geological memory lane consisting of the following scenarios:

- A *snowball earth:* Ice covered the entire planet, making it practically uninhabitable until around 635 million years ago.
- A *greenhouse earth:* The climate was tropical, including at the poles. During this period, which included the age of the dinosaurs, global temperatures were 7.2°F to 12.6°F (4°C to 7°C), perhaps 18°F (10°C), warmer than today. Furthermore, CO_2 concentrations six times preindustrial levels were common. This era lasted until about 70 million years ago.
- An *icehouse earth.* Starting around 34 million years ago, this period consisted of glacial-interglacial pulses. Over the course of the last 2.6 million years of this era, ice sheets formed over the European and North American land masses, pulsing every 40,000 years. The Pleistocene lasted from 1.8 million years ago until 11,550 years ago.

- A *Holocene interglacial:* This began at end of the Pleistocene.

While we as a species hopefully will be around for some time to come, the Potsdam Institute of Climate Impact Research's timetable for decline indicates that ultimately we will become a mere cosmic memory. According to the institute's sobering timetable, in 800 million years, the average global temperature will have risen to 54°F (30°C); in 1.2 billion years it will have risen to 22.2°F (40°C); and in 1.6 billion years, it will have reached 126°F (70°C), making photosynthesis and life as we know it impossible. In somewhere between 3.5 and 6 billion years, the sun will have grown to the point that temperatures on Earth will exceed 1,800°F (1,000°C), resulting in the disappearance of the atmosphere and the melting of rocks (see Behringer 2010:14).

Over the course of their some 5 to 6 million years on the face of the planet Earth, humans have been described by some as the "children of the ice" (Behringer 2010:39). The Earth has experienced 10 major and 40 minor episodes of glaciations over the past 1 million years (Farley 2008:78). Milankovitch cycles in which the tilt of the Earth's axis fluctuates between 22 and 24.5 degrees about every 41,000 years cause the beginning and ending of ice ages. While the sun may contribute to climate change, according to John Farley (2008:79), an academic in the department of physics and astronomy at the University of Nevada, Las Vegas, "in the last quarter of the twentieth century, solar changes can account for less than one third of the observed warming." The climate, for the better part of the past 110,000 years, has fluctuated between "warm" states resembling the present time and regime and prolonged "cold" states marked by glacial advances and temperatures of 8°C or more below the present average, with the Last Glacial Maximum occurring about 20,000 years ago (Kennedy 2006:47).

Human societies began to make the transition from foraging or hunting-and-gathering societies to horticultural village societies about 10,000 years ago and the transition to stratified state societies starting about 6,000 years ago. These transitions have occurred in the context of what geologists call the Holocene, generally believed to be an interglacial period that has seen only minor shifts in climate, such as the Medieval Warm Period (AD 950 to 1250) and Little Ice Age (AD 1300 to 1850). Climate change, although probably more of a naturally induced form rather than an anthropogenic one, has played a role in the collapse of ancient civilizations, such as the Classic Maya in the ninth century at the time of the most severe drought experienced during the first millennium (Diamond 2005; Fagan 2008).

Current forces that can affect climate include (1) changes in the sun's energy output; (2) variations in the distance of the Earth from the sun and in the angles at which solar radiation reaches various parts of the Earth; (3) changes in the atmospheric and oceanic circulation systems; (4) changes in the absorption or radiation of energy by the Earth's surface, related to the extent of the cloud cover and the nature of the surface; (5) possibly volcanoes; and (6) the greenhouse effect (Farley 2008:69; Officer and Page 2009:109). All of these except for the last are natural. Atmospheric CO_2 hovered between 180 and 300 ppm over the course of the 650,000 years before recent times (Maslin 2009:8). CO_2 hovered around 280 ppm during the last 10,000 years until the onset of the Industrial Revolution.

Anthropogenic Climate Change

In the Northern Hemisphere the average temperature rose about 1°F (1.8°C) from 1900, declined 0.5°F (0.9°C) between 1940 and 1970, then began to increase rapidly again (Officer and Page 2009:107). Indeed, many climate scientists in the early 1960s viewed this temporary cooling phase as the onset of the next ice age. In time, it became apparent that this period of global cooling was a result of *global dimming*, resulting from a variety of anthropogenic activities that hindered sunlight's reaching the Earth's surface. These anthropogenic activities included the effects of urbanization and manufacturing and increased motor vehicle and aircraft exhaust. According to Ruddiman (2005:172), "Industrial-era emissions of sulphate aerosols have probably cancelled part of the warming that greenhouse-gas emissions would have otherwise caused." Furthermore, the decline in the intensity of sunspots in the 1960s and 1970s contributed to a cooling trend (Maslin 2009:211). Even today, global dimming may be occurring in various places, such as China, that have embarked upon paths of intense industrialization that "account for local cooling by reflecting considerable amounts of solar radiation back into the atmosphere" (Luke 2008:125).

The temperature in the Southern Hemisphere has risen steadily by 1°F (1.8°C) over the course of the past 100 years. In 1971 the Stockholm Study of Man's Impact on Climate warned that humanity faced the risk of future climatic shocks (Weart 2003:100). Michael Mann, Raymond S. Bradley, and Malcolm K. Hughes in 1998 conducted a study on global warming that claimed that, on the whole, the 1990s had been hotter than any other decade during the previous six centuries and that this warming was largely due to anthropogenic activities (Behringer 2010:3). They portrayed the climate curve of last 1,000 years in the form of a hockey stick, with not much happening for 900 years until temperatures steeply increased in the late twentieth century. The current rate of warming has been about 10 times faster than any rate in the past 10,000 years (IPCC 2007). Furthermore, CO_2 atmospheric concentrations are significantly higher than at any time in the past 400,000 years, during most of which CO_2 concentrations cycled between 180 and 300 ppm, followed by an increase over the past century or so from 280 to 390 ppm. Unless drastic steps are taken, the atmospheric CO_2 level will continue to rise rapidly during the course of the twenty-first century.

The 2007 Intergovernmental Panel on Climate Change (IPCC) synthesis report states that there was a 0.75°C increase in global temperatures and a 22-centimeter increase in sea levels over the course of the twentieth century. It also predicts that global temperatures could increase further, by between 1.1°C and 6.4°C by 2100, and sea levels could rise between 28 and 79 centimeters by 2100—even more if the melting of Greenland and Antarctica accelerates. Australian atmospheric scientist A. Barrie Pittock (2008:19) argues that given the uncertainties in climate science, "many scientists have consciously or unconsciously downplayed the more extreme possibilities at the high end of the uncertainty range in an attempt to appear moderate and 'responsible' (that is, to avoid scaring people)." *The Copenhagen Diagnosis*, a report that seeks to synthesize most policy-relevant climate science published since the 2007 IPCC report, was released in time for the UN Copenhagen conference in December 2009 (Allison et al. 2009:5). The report indicates that 2008 constituted the ninth warmest year on record, one in which La Niña caused a temporary dip in average global temperatures (Allison et al. 2009:11). Despite the fact that the sun exhibited extremely low brightness over the course of the previous three years (Allison et al. 2009:13), numerous temperature records had been broken during this period. Years 2007, 2008, and 2009 saw the lowest summer Arctic sea ice cover ever recorded. The Northwest Passage and Northeast Passage simultaneously were ice-free for first time in 2008, a phenomenon repeated in 2009. Every single year of the twenty-first century has been among the top 10 warmest years since instrumental records began, with winters warming faster than summers (Allison et al. 2009:14). Continuing marked increases in hot extremes and decreases in cold extremes are expected in most areas across world (Allison et al. 2009:15). *The Copenhagen Diagnosis* reports that the mean global temperature is expected to increase 7.2°F to 12.6°F (4°C to 7°C) by 2100 (Allison et al. 2009:49).

Warming is having the following effects:

- The cryosphere is losing ice at an unusually rapid rate, with the rapid and general retreat of glaciers, shrinkage of the annual Arctic sea ice, and collapse of ice shelves.
- The oceans are warming, becoming more saline, rising, and absorbing less CO_2.
- Animal species are retreating to higher altitudes and latitudes.
- The rise in global average temperatures is unprecedented in the last 2,000 years.
- The most profound warming is occurring at the poles, with the Northern Hemisphere leading the Southern Hemisphere.
- Measurements of solar irradiance suggest that the Earth's climate should be getting cooler rather than warmer (Morgan and McCrystal 2009:242–43)

James Hansen, the provocative director of the NASA Goddard Institute for Space Studies (GISS) based at Columbia University, predicted at a US Senate hearing in 1988 that warming would be greatest at the poles and that perhaps in three decades the Arctic Ocean would be ice-free in the summer months (Officer and Page 2009:199). He spoke out again in the Senate and in online commentary in 2008 about the drastic state of the Earth's climate system, noting that the Arctic ice has been melting at a faster rate than predicted previously and that the North Pole will be ice-free by the end of Arctic summer 2030. Hansen regards the melting of the Arctic ice cap as an alarming tipping point and argues that failure to act could result in mass extinctions. He maintains that humanity needs to reduce greenhouse gas emissions to the 1988 level, which was 350 CO_2 ppm.

Global and regional temperatures during any period exhibit a certain amount of variability. For example, the United Kingdom experienced relatively cool weather in 2008, with a wet July and August (Lovelock 2009:2). The surface water saw a cooling in the Gulf of Mexico. The Arctic regained a little of its ice in the wake of the astounding losses of 2007, although ominously ice continues to grow thinner. As Lovelock (2009:2) observes, climate change is "rarely smooth: it goes by fits and starts."

Greenhouse Gas Emissions

The principal greenhouse gases include carbon dioxide, nitrous oxide, methane, water vapor, the chlorofluorocarbons, and ozone. Table 1.1 depicts the global warming potential of selected greenhouse gases.

Carbon Dioxide

Carbon dioxide comes mainly from the burning of fossil fuels, deforestation, destruction of carbon-rich soils, and production of cement from limestone. Table 1.2 depicts the atmospheric concentrations of carbon dioxide from 1960 to 2005.

Current atmospheric CO_2 levels are higher than they have been in the last million years. Global CO_2 emissions have been

Table 1.1. Global Warming Potential of Selected Greenhouse Gases

Greenhouse Gas	*Global Warming Potential**
Carbon dioxide	1
Methane	25
Nitrous oxide	298
Hydrofluorocarbons	124–14,800
Perfluorocarbons	2,390–12,200
Sulfur hexafluoride	22,800

*Global warming potential refers the heat-trapping power of a greenhouse gas relative to CO_2 over a 100-year time frame.

Source: Adapted from McKeown and Gardner (2009).

growing at about 3 percent per year since 2000. Global emissions of CO_2 from fossil fuel combustion and cement production rose from 22.6 billion tons in 1990 to 31 billion tons in 2008, a 37 percent increase (Flavin and Engelman 2009:7). US CO_2 emissions from fossil fuel combustion grew by 27 percent between 1990 and 2008, and in China they grew by an astounding 150 percent, from 2.3 billion to 5.9 billion tons. While Russia, which underwent tremendous deindustrialization in the wake of the collapse of the Soviet system, saw a fall in emissions of one-third between 1990 and 2005, China and India have more than doubled their emissions since 1990. Conversely, total greenhouse gas emissions appear to have dropped in 2009 due to the Global Financial Crisis (Allison et al. 2009:9). CO_2 that has not been absorbed by

Table 1.2. Atmospheric Concentrations of Carbon Dioxide, 1960–2005

Year	*CO (ppm)*	*Emissions (billion tons of carbon)*	*Temperature (°C)*
1960	316.91	2.53	13.99
1970	325.68	4.00	14.03
1980	338.68	5.21	14.18
1990	354.19	5.99	14.38
2000	369.48	6.45	14.33
2005	379.66	7.56	14.63

Source: Adapted from Sawin (2008:42–43).

the oceans, trees, or other means has a "removal time of more than 100 years, perhaps as long as 1000 years" (Richter 2010:21).

Methane

Although methane (CH_4) has a removal time of about 10 years, it is 64 times more powerful than CO_2 in terms of climate change potential over 20 years and 23 times more powerful over 100 years. Methane comes from biomass decomposition, coal mining, natural gas and oil system leakages, livestock production, waste water treatment, landfills, rice cultivation, burning of savannah, and burning of fossil fuels. Given problems with measuring methane levels in the atmosphere, some scientists contend that its impact generally has been underestimated. With rising temperatures, there is the danger that the methane locked up in permafrost will be released as it hydrates in the oceans.

Nitrous Oxide

Nitrous oxide (N_2O) comes from the heavy use of nitrogen fertilizers in industrial agriculture, the production of synthetic materials, and the burning of fossil fuels. It is 296 times more powerful than CO_2 over a 100-year period and remains around for 120 to 150 years.

Water Vapor

Global warming results in the evaporation of water from the oceans; this vapor can turn into clouds, which shade the Earth during the day but keep it warm at night. Clouds, of course, may release rain and thus disappear. Clouds can absorb heat radiation from the sun but can also reflect sunlight.

Flurocarbons

The flurocarbon (F-gas) family consists of chloroflurocarbons (CFCs), hydroflurocarbons (HFCs), perfluorocarbons (PFCs), and sulfur hexafluoride (SF_6). Most of these gases come from refrigeration and air-conditioning, including in cars. They are used as solvents, as blowing agents in foams, in aerosols or propellants, and in fire extinguishers. The F-gases were developed by the chemical industry and could be generally replaced by environmentally safe, efficient, and technologically proven alternatives. The fluorocarbons have a lifetime of about 1,000 years (Richter 2010:24).

Impacts of Climate Change on the Environment

In terms of its impact on the environment, climate change has contributed to, and will continue to contribute to, a rising average global temperature; rising sea levels; warming oceans; melting ice caps, glaciers, and tundras; erratic weather patterns; and the loss of biodiversity. Table 1.3 depicts some environmental risks at different global temperature increases above the preindustrial level.

Table 1.3. Some Potential Environmental Risks at Different Global Temperature Increases

Temperature Rise	*Environmental Impacts*
1°C	Weakening of Atlantic thermohaline circulation; thawing of permafrost; continuing retreat of glaciers; at least 10 percent of land species facing extinction; bleaching of 80 percent of coral reefs, including Great Barrier Reef
2°C–3°C	Potential for Greenland ice sheet to begin melting, thus increasing sea level rise to seven meters or more; increased risk of collapse of West Antarctic Ice Sheet; heightened risk of collapse of Atlantic thermohaline circulation; possible extinction of 15 to 50 percent of Arctic species; more coastal flooding
4°C	Loss of half of the Arctic
5°C	Possible disappearance of large Himalayan glaciers and increased ocean acidity, impacting adversely on marine ecosystems

Source: Adapted from Jarman (2007:10–13).

Rising Temperatures

The Earth's overall surface temperature has increased 1.368° F (0.76°C) plus or minus 0.34°F (0.19°C) since 1850 (Camilleri and Falk 2010:258). The global temperature curve seems to have had a general downturn from 1940 until the mid-1970s (Maslin 2009:209–10). Some climate scientists argue that this downturn was caused by global dimming related to an increased output in the atmosphere from both volcanoes and industrial activity, such as the burning of coal, oil, and wood, as well as tiny airborne particles of soot, ash, sulfur compounds, and other pollutants (Victor 2004:10). Aerosol particles and other particulates given off by industrial activities absorb solar radiation and reflect it back into space. Other sources of global dimming include the burning of tropical grasslands and forests and desert dust storms. Furthermore, the cooling trend of the 1960s and 1970s appears to have been due in part to the influence of the sunspot cycle (the intensity of sunspots varies over 11-year periods). Various climate scientists argue that global dimming produces a cooling effect that may have partially masked the impact of greenhouse gases on global warming. According to Robert Henson (2006:182), "Most of the world's highly industrialized nations began clearing up their smokestacks and tailpipes by the 1970s, and the economic downturn of the 1990s across the Eastern bloc reduced aerosol production." Fred Pearce (2007:110) reports,

> Since the fall of the Berlin Wall, the old polluting industries have been mostly shut down, and the air has cleared. More sun penetrates the smog-filled landscape, and central Europe has been warming correspondingly. In the past fifteen years, temperatures there have risen three times the global average rate.

Rapid industrialization in China, India, and other developing countries has contributed to global dimming over the past several decades while at the same time contributing to greenhouse gas emissions. Ironically, while most the globe has been experiencing increased temperatures, various parts of the world with highly dense populations have experienced declining temperatures (Pearce 2007:109). Some might argue that global dimming serves to curtail climate change and global warming, at least in the short term, but due to the impact of aerosols and particulates in terms of quality of air, depletion of the ozone layer, and human health, this hardly constitutes a justifiable rationale for allowing their ongoing emission into the atmosphere.

Pearce (2010:17), however, summarizes another take on the lowering of global temperatures between the 1940s and 1970s:

> Most climate scientists now agree that the cold decades from the 1940s to 1970s had little to do with either man-made pollution or planetary wobbles. The mid-century cooling was most associated with two natural phenomena: first the eruption of a cluster of medium-sized volcanoes that pumped sunlight-scattering sulphate particles into the upper air, and second oscillations such as the Pacific Decadal Oscillation, a kind of slow-motion El Nino that moved heat out of the atmosphere into the oceans.

The global annual mean temperature began to rise again in the late 1970s. NASA's GISS reports that temperatures began to climb in 1977 and have been above the norm every year since. The twentieth century was the warmest century of the past millennium, and the period of 1990 to 2000 was the warmest decade of the past millennium. GISS reports that 2005 and 2010 were tied for the status of the warmest year ever on record. Conversely, the World Meteorological Organization and the UK Climate Research Unit at the University of East Anglia reported that 2005 was the second-warmest year, slightly behind 1998. These differences reflect varying ways of measuring global temperature, but ultimately the conclusion is the same: the planet is growing ever warmer in measurable and consequential ways.

Table 1.4 depicts the 10 warmest years on record from 1880 to 2008. Worldwide 2010 was the wettest year on record but also a very hot year. The World Meteorological Organization maintained that 2010 was the hottest year since records began in 1850, and NASA and the National Ocean and Atmospheric Administration also reported that 2010 was the wettest year on

Table 1.4. Ten Warmest Years on Record, 1880–2008

Ranking	*Year*
1	2005
2	1998
3	2002
4	2003
5	2006
6	2007
7	2004
8	2001
9	2008
10	1997

Source: McKeown and Gardner (2009).

record as well as a very hot year, tying with 2005 as the hottest year on record.

Temperature increases vary considerably around the planet. For example, a global temperature rise of 5.4°F (3°C) would translate into only a 1.8°F to 3.6°F (1°C to 2°C) increase over most of the oceans but a rise of 12.8°F to 14.4°F (7.1°C to 8°C) in the Arctic (Paskal 2010:68). Gunter Weller, the director of the Center for Global Change and Arctic System Research at the University of Alaska, Fairbanks, reports that mean temperatures in Alaska increased by 5°F (2.74°C) in summer and 10°F (5.45°C) in winter over the past three decades or so (Johansen 2006:299). Alaska has become one of the fastest-warming regions on the planet. The Arctic Climate Impact Assessment team of some 300 scientists and indigenous peoples reports that annual temperatures have risen by 5°F (2.74°C) in Alaska since the 1960s and that average winter temperatures increased 8°F during that period (Kister 2005:3). Southern Alaska has been experiencing salmon kills due to increased temperatures (Kister 2005:20).

Rather than drought, which had been plaguing much of Australia for over a decade up until 2009, much of eastern Australia was severely impacted by heavy rains, hurricanes, and floods during late 2010 and early 2011. Australia was ravaged by one of the most severe La Niñas that it had experienced in recorded history. David Jones, an Australian Government Bureau of Meteorology analyst, observed, "The last year of extreme weather events has been really extreme, but in the Australian context the really major story is La Nina" (quoted in Tippet and Russell 2011:9). The recent La Niña has been aggravated by a record-high sea surface temperature, which very likely is related to climate change. Ironically, Australia in 2010 experienced its coolest year since 2001, but it was still warmer than the 1961–1990 average. The Australian Government Bureau of Meteorology (2010:v) reported,

> The last decade (2001–2010) was the warmest ten-year-period on record (0.52°C above the average). Records indicate that Australia's climate has steadily warmed over the last 60 years, with very few cool years occurring in the last three decades.

The IPCC predicts that average global temperatures will increase by 4.5°F (2.5°C) by 2100. In contrast, the Massachusetts Institute of Technology's Joint Program on the Science and Policy of Global Change predicted in April 2011, based on highly comprehensive modeling of climate change impacts, a temperature rise over 9°F (5°C) by 2100 (Primm and Reilly 2011:3). Based on the 2007 IPCC projection, table 1.5 depicts global average surface temperature increases by 2100 under various scenarios.

According to NASA, the hottest temperature on record in Asia occurred in Pakistan, when the temperature hit 53°C in July 2010. Beijing had its hottest day on record—namely, 40.6°C—in August 2010. Moscow experienced temperatures of up to 40°C during summer 2010. On July 11, 2010, the previous hottest temperature in Russia of 43.8°C, set on August 6, 1940, was broken by a temperature of 44.0°C in the European portion of the country near the Kazakhstan border (Jeff Masters's WunderBlog, July 19, 2010, www.wunderground.com). NASA reported that a record-breaking heat wave in western Russia during summer 2010 was more a swan song occurrence in that it was "well beyond the normal expectations in the instrumental record" and was very likely the result of an "extreme pattern of atmospheric

Table 1.5. Projected Global Surface Warming by 2100

	Global Average Temperature Changes Relative to 1980–1999			
	Best Estimate		*Range*	
Scenario	*°F*	*°C*	*°F*	*°C*
B1	3.2	1.8	2.0–5.2	1.1–2.9
A1T	4.3	2.4	2.5–6.8	1.4–3.8
B2	4.3	2.4	2.5–6.8	1.4–3.8
A1B	5.0	2.8	3.1–7.9	1.7–4.4
A2	6.1	3.4	3.6–9.7	2.0–5.4
A1F1	7.2	4.0	4.3–11.5	2.4–6.4

Source: Adapted from Richter (2010:42). The IPCC scenarios are as follows: The A1 scenarios assume rapid economic and population growth with A1F1 entailing reliance on fossil fuels; A1T, nonfossil energy; and A1B, a combination. The B1 and B2 scenarios entail some mitigation of greenhouse gas emissions, through increased energy efficiency and technological improvement (B1) and through more localized solutions (B2).

winds—widely referred to as blocking," which could not be accounted for by anthropogenic climate change (NASA, The Russian heat wave of 2010). However, NASA conceded that it is not clear whether greenhouse gas emissions may influence the frequency or intensity of wind blocking during summers (NASA, The Russian Heat Wave of 2010, 3). James Hansen at NASA, however, contends that the recent Russian heat wave would probably not have occurred if CO_2 levels had remained below preindustrial levels (NASA GISS 2010). While 2010 was globally a very hot year, Europe and parts of North America experienced unusually heavy snowfalls and cold weather and eastern Australia experienced relatively cool weather and heavy rainfall in late 2010 due to La Niña, which in turn was driven by a warming of the western Pacific Ocean.

Rising Sea Levels, Warming Oceans, Melting Ice Caps, Glaciers and Tundras, and Disintegrating Peat Bogs

The IPCC (2007) reports that due to a rise in the average global surface temperature, sea levels around the world increased an average of 1.8 millimeters per year between 1961 and 2003. The IPCC projects a sea level rise of another 18 to 59 centimeters over the course of the present century. Sea levels are rising in a warming world, not only as a result of the addition of all the water locked up in ice sheets and glaciers but also because water expands when warmed. The oceans absorb perhaps 40 percent of CO_2 emissions from the burning of fossil fuels (Hossay 2006:8). Over time, the warming oceans may absorb less CO_2, thus even further warming the planet. There has been an increase in the size and number of oceanic "dead zones" with low levels of oxygen, which are by and large devoid of living organisms and concentrated along the continental shelves of highly populated areas (Hetherington and Reid 2010:271).

Climate change may be causing stronger and more frequent El Niños (Pearce 2006:36). The melting of glacier and the projected melting of portions of the Greenland and Antarctic ice caps will also contribute significantly to the rise of sea levels. A NASA study in late 2002 found that the ice pack covering the Arctic Ocean is vanishing at the rate of about 9 percent per decade (Gelbspan 2004:21). Between 2005 and 2009, Greenland lost between 380 and 490 tons of ice, about 150 billion tons more than it acquires in snowfall each winter (Maslin 2009:109). It is important to note, however, that the melting of the Arctic Ocean ice does not contribute to making sea levels rise. James Hansen, the director of NASA's GISS, asserts,

> The broader picture gives a strong indication that ice sheets will respond in a non-linear fashion to global warming . . . and are already beginning to do so. There is enough information now, in my opinion, to make it a near certainty that business-as-usual scenarios will lead to disastrous multi-metre sea level rise on the century time scale. (Quoted in Camilleri and Falk 2010:294)

Various climate scientists suggest a rise of between 0.75 and 1.9 meters by 2100 if emissions continue to follow the present trajectory (Chivers 2009:49).

Some experts predict ice-free summers in the Arctic by 2030, if not earlier (Taylor 2008:35; Hansen and Hoffman 2011:13).

Indeed, "for the first time in recorded history a hole large enough to be seen from space opened in the sea ice above the North Pole" in 2000 (Maslin 2009:53). Melting ice will leave a greater absorptive surface that will reflect less heat back into the atmosphere. Furthermore, the melting of the permafrost could release large quantities of trapped methane, a process that has already started. In Antarctica 10 major ice shelves have collapsed or retreated since 1980 (Morgan and McCrystal 2009:128).

Most glaciers around the world are retreating, in the Alps, Kenya and other parts of Africa, Papua New Guinea, the Andes, North America, China, the Himalayas, and New Zealand (Pittock 2008:5). Glacier National Park in Montana contained over 150 glaciers in the late 1800s and has only about 35 left; it is predicted that it will have no glaciers left by 2030 (Diamond 2005). The glacier on Mount Kilimanjaro in East Africa underwent a decline of about 80 percent between 1912 and 2000, and 95 percent of Alaskan glaciers have experienced a doubling of their thinning or reduction rate since the mid-1990s (Lynas 2004:218). The Columbia Glacier in Alaska reportedly retreated nearly 13 kilometers (8.1 miles) between 1982 and 2000 (Kister 2005:26). Glaciers in the Himalayas and on the Tibetan Plateau are rapidly retreating. Yao Tandong at the Chinese Academy of Sciences claims that glaciers on the Tibetan Plateau "[have] been shrinking at the rate of four metres a year since he started monitoring them in 1989" and predicts 40 percent of the glaciers on the plateau will have disappeared by 2050 (Watts 2010:48). The Gangetic glacier, the source of the Ganges River, is retreating about 30 meters every year (Shiva 2008:11). The annual summer melt of the Hindu Kush and Himalayan glaciers is a major water source for China, India, Pakistan, and much of continental Asia. Increased melting would cause greater flow for several decades, after which some heavily populated regions will likely run out of water. Glaciologists have discovered massive river systems of melt water beneath the ice sheets of Greenland and Antarctica (Pearce 2006:27).

The melting of the Arctic tundra is expected to release massive amounts of methane into the atmosphere. In recent years northern ponds and marshes have remained unfrozen even during winter due to methane emissions. An estimated 1 million square kilometers of the Siberian permafrost has thawed 60 percent since the 1970s, and the rate of methane emissions in Alaska has even been higher due to higher temperatures (Dawson and Spannagle 2009:273). Rising temperatures are contributing to the thawing of peat bogs, which are wetlands located in cold, temperate areas primarily in the Northern Hemisphere, resulting in the release of methane into the atmosphere. The West Siberian bogs cover over 600,000 square kilometers. Other peat bogs are situated in Ireland, northern Germany, Scandinavia, Canada, Alaska, and the northern regions of Minnesota and Michigan in the United States.

Erratic Weather Patterns

Climate change may contribute to the intensification of droughts, hurricanes, cyclones, precipitation, and flooding. Droughts have become commonplace in recent decades in places as diverse as the American Southwest, the Sahel in sub-Saharan Africa, the Amazon Basin, and much of Australia. Bushfires have increased in frequency and intensity in much of the American Southwest, Mediterranean Europe, and southeastern and southwestern Australia. Furthermore, as Lara Hansen and Jennifer Hoffman (2011:12) observe,

> Increasingly fire frequency and intensity are also likely to speed the rate of climate change both locally and globally. A large, hot fire may release in hours carbon that it has taken decades for a forest to restore, instantly increasing the amount of carbon dioxide in the atmosphere and decreasing the ecosystem's capacity to take up new carbon.

As temperatures rise, more water tends to evaporate from the ocean, which means that overall more moisture is available in the atmosphere to produce rain. Central India has experienced rain events more extreme in number and intensity since the mid-twentieth century. Between 1951 and 2000, there was a doubling of rainstorms producing in excess of 10 centimeters in one day (Shiva 2008:11). Climate change over the course of the present century is expected to result in the following scenarios: (1) increasing precipitation in higher latitudes, leading to increased winter and spring runoff and flooding in some areas; (2) decreasing precipitation and increasing drought frequencies in lower latitudes; (3) increased summer evaporation and decreasing surface flow and soil moisture in mid- to high-level latitudes; (4) decreasing lake levels in some areas, with changes in wetland communities; and (5) decreasing per capita water availability, particularly in low-latitude countries with high population growth rates. Per capita water availability in Africa has decreased by 75 percent over the course of the past half century (Hardy 2003:81).

While much of the Earth is warming, some regions may experience a cooling effect as a result of warming ocean waters. There is evidence that the North Atlantic Conveyor Belt, of which the Gulf Stream is a part, is slowing down and may be contributing to recent severe winters in northwestern Europe.

Loss of Biodiversity

Because of climate change, plants and animals are moving into regions closer to the poles because these regions are becoming warmer. Mammals, birds, butterflies, fish, and insects are moving toward the poles and higher elevations in order to survive. Animals, such as polar bears, that have adapted to Arctic conditions cannot migrate and are in danger of extinction due to the contraction of the Arctic ice pack and a reduction in the populations of animals, such as seals, upon which they feed. Indeed, polar bears have been venturing onto land as the sea ice increasingly melts during the summer months, and some of them have been interbreeding with grizzly bears, resulting in a new type of bear called a "pizzly" or "grolar bear" (Barnosky 2009:10–14). Since 1992, the Kenai Peninsula in Alaska has experienced an infestation of bark beetles due to higher temperatures (Hillman 2004:25). Coral reef damage is occurring all over the world, including at the Great Barrier Reef off the northeastern coast of Australia (Henson 2006:117–20).

Safe Temperatures and Tipping Points

Numerous climate scientists have attempted to define a "safe temperature" limit. The Hadley Centre in the United Kingdom provides the scenarios presented in table 1.6 regarding the impact of various emissions levels on global temperatures.

The Copenhagen Diagnosis asserts that global greenhouse gas emissions need to peak between 2015 and 2020, then decline rapidly, if global warming is to be limited to a maximum of 2°C above preindustrial values (Allison et al. 2009:7). Stabilizing CO_2 emissions at 445 ppm would require a drop of 89 percent in global emissions. At 445 ppm, global temperatures would still rise by 2°C (relative to preindustrial times) (Li 2008:60). Christopher Shaw (2010) queries the 2°C limit that numerous governments and corporations, the European Union, and even many NGOs have adopted as the safe temperature limit. He argues that this arbitrarily designated limit "makes climate change a problem for the future which allows humanity to continue with 'business as usual' whilst the search for a techno-fix continues," an argument that anticipates my critique of existing climate regimes and green capitalism with its emphasis on ecological modernization, which I discuss later in this book.

Table 1.6. Concentration Levels and Temperature Levels

Stabilization Level (ppm CO_2)	*2°C*	*3°C*	*4°C*	*5°C*	*6°C*	*7°C*
450	78	18	3	1	0	0
500	96	44	11	3	1	0
550	99	69	24	7	2	1
650	100	94	58	24	9	4
750	100	99	82	47	22	9

Source: Adapted from Stern (2009:26).

The IPCC distinguishes between Type I climate change, which is gradual, and Type II, which is much more abrupt and results the crossing of *critical tipping points*. Pearce (2006:346) asserts that both humanity and the planet may be entering a "terra incognita climatically" that is manifesting itself in melting Arctic ice, the possible collapse of the West Antarctic Ice Sheet, the contraction and demise of the Amazon Rainforest, the acidification of the ocean, and the increasing emission of methane from a number of sources, including peat bogs.

Another profound danger posed by reaching tipping points includes the risk of a complete shutdown in the circulation of the major Atlantic Ocean currents, resulting in a drastic cooling of Europe. The thermohaline circulation that drives the oceans' conveyer belt is endangered as cold freshwater from melting ice from the Arctic and Greenland ice caps hits the Atlantic Ocean. James Hansen and his colleagues at GISS argued in a 2007 paper for a limit of 1.7°C on the basis that potential changes above this level—including irreversible loss of Greenland and Antarctic ice sheets and species extinction—would be "highly disruptive" (Hare 2009:19). Hansen maintains that humanity needs to reduce atmospheric carbon levels below the present 390 ppm, to 350 ppm or less, in order to avoid irrevocable damage to human societies and the planet. If a tipping point is passed, then a subsequent cooling of the climate system would not necessarily reverse the change. Peter Ward (2007:146) notes that research on ice-core records indicates that the average global temperature around 200,000 years ago and again 10,000 years ago shifted as much as 18°F (10°C) in a matter of a few decades:

> The current *average* global temperature is 59 degrees Fahrenheit [32.4 degrees Centigrade]. Imagine that it suddenly shot to 75 degrees Fahrenheit [41 degrees Centigrade] or dropped to 40 degrees Fahrenheit [22.2 degrees Centigrade], in a few decades. We have no experience of such a world and what it would be like; such sudden perturbations in temperature would enormously alter the atmospheric circulation patterns, the great gyres that redistribute Earth's heat. At a minimum, such changes would create catastrophic storms of unbelievable magnitude and fury.

The Impact of Climate Change on Human Societies

Climate change threatens to have serious impacts upon many of the peoples of the world, including foragers, horticulturalists, pastoralists, peasants, industrial farmers, and city dwellers. The Global Commons Institute estimates that damage resulting from climate change may reach US$400 billion by 2012 and an astronomical US$20 trillion by 2050 (Hossay 2006:13).

Settlement Patterns

Climate change endangers people's cultures as a result of rising temperatures and sea levels, droughts, heavy rains, hurricanes, and cyclones. A sea level rise of 0.2 to 0.7 meters could result in increased beach erosion and coastal flooding, the loss of various coastal ecosystems (such as mangroves, wetlands, the Great Barrier Reef), the displacement of millions of people from low-lying areas, and salt water intrusion into coastal aquifer water supplies, posing a danger to farming and human habitation. Rising sea levels threaten entire populations of islands, particularly in the South Pacific and approximately 1,200 islands of the Maldives in the Indian Ocean. Tuvalu in the South Pacific is often depicted as a "canary in the coal mine" for climate change, and New Zealand, while not officially recognizing the category of "environmental refugee" or "climate refugee," has increased its intake of Tuvaluans, but only if they are employable and under age 45. Over 300 million people live within three feet of sea level (*National Geographic* 2004:28). There are presently some 10 million environmental refugees in the world, and it has been estimated that there could be 150 million by 2050 (Cowie 1998). Coastal megacities in danger of flooding include Shanghai, Calcutta, Lagos, London, Rotterdam, New York, Miami, and New Orleans. Populations under threat from a rise in sea level include many people living in Vietnam, Bangladesh, eastern China, India, Thailand, the Philippines, Indonesia, and Egypt. The Native Alaskan village of Shishmaref, home to some 600 Native Alaskans living at the far western edge of the state about 60 miles north of Nome, has been eroding into the Bering Sea due to rising seas and increased storm surges (Johansen 2006:308).

Subsistence and Food Security

The Arctic Climate Impact Assessment report sponsored by nations with an interest in the region pays special attention to species important to the Arctic's indigenous peoples (Symon, Arris, and Heal 2004). Indigenous Arctic peoples will have to cope with the loss of sea ice for hunting and fishing, changed animal migration patterns, loss of permafrost, and changes in the availability of traditional food sources. The Inuit of western Greenland and the Canadian Arctic islands, who hunt Peary caribou during the summer months, have experienced a decline in these herds from 26,000 in 1961 to 1,000 in 1997 (Flannery 2005:100). Other mammals on which the Inuit have relied, including polar bears, seals, and walruses, have also become endangered species due to climate change. Huslia, an Athabaskan village some 300 kilometers (186 miles) west of Fairbanks, has experienced the disappearance of nearby lakes that provided valuable food resources for them (Lynas 2004:52).

The UN Food and Agriculture Organization has warned that in some 40 percent of the poorest developing societies with some 2 billion inhabitants, climate change could dramatically increase the numbers of malnourished people (Monbiot 2006:6–7). There is evidence that droughts in various sub-Saharan countries may be related to climate change. Anthropologist Megan Jennaway (2006:11–12) presents a grim picture of the possible impact of climate change on East Timor:

> One of the first impacts will be a shifting of the coastline and a loss of estuarine breeding grounds for fish and bird populations, particularly mangrove, seagrass and coral habitats. . . . Overall rainfall is predicted to decrease significantly over the next 50 years, with all the attendant impacts upon crop production and local forest ecologies. There will also be an increase in the number of El Nino events, with more frequent and longer droughts becoming the norm. Taking all of this into

> account, it may mean that East Timor's traditional ways of life, with a majority of its people living in rural areas and engaged in intensive modes of agricultural production, may well vanish beneath an inexorable tide of runaway global warming.

Developed countries, which obtain many agricultural products from the developing world, also stand to suffer from climate change. The Swiss village of Saas-Balen consists of 423 inhabitants who reside three kilometers below the Gruben glacier, which has been melting for over a century and has been losing 60 to 70 centimeters in height annually (Cowie 1998:273). The village was flooded in 1968 and again in 1970, eventually prompting its residents to drain one of the melt-water lakes in 1995 to prevent further flooding. Wilfried Haebrli, director of the United Nations' Glacier Monitoring Service, linked the threat of the village being flooded by mud and water to climate change (Cowie 1998:273). Montana in the northwestern United States had been adversely affected by climate change: a state that has historically experienced marginal rainfall has become even drier. Drought has resulted in the abandonment of many farms in eastern Montana (Diamond 2005:49).

According to UN figures, about half of the world's population relies upon mountain-produced water for agriculture, electricity, industrial production, and drinking purposes (Lynas 2004:235). Andean mountain villages and towns in Peru, Ecuador, Bolivia, and Colombia are losing water for both irrigation and drinking. La Paz and Quito both derive their water supply from glacial runoff, which may eventually diminish. Mountains in humid regions supply an estimated 30 to 60 percent of downstream freshwater. Melting of the Sierra ice pack will increase the likelihood of water shortages in Los Angeles (Diamond 2005:502).

As Tim Flannery (2005:204) observes, cities "constitute fragile entities vulnerable to stress brought about by climate change." Major cities in the developed world already suffering from water shortages, possibly related in part to climate change, include New York, Los Angeles, Chicago, Washington, DC, Tucson, Sydney, and Melbourne (Glantz 2003:58).

The Impact of Climate Change on Health

Various scholars have recognized the impact of climate change on health. In his now classic *Planetary Overload*, Tony McMichael (1993), an epidemiologist at the Australian National University, discusses the direct effects of global warming on health in the form of heat stress and respiratory ailments, as well as the indirect effects in terms of the spread of vector-borne and water-borne diseases. More frequent heat waves, particularly in urban areas, threaten the health and lives of vulnerable populations, such as the elderly, the sick, and infants. Indeed, cities act as "heat islands" due to the presence of concrete roads, buildings, factories, and motor vehicle exhaust fumes. The mortality of some 35,000 people during the heat wave of summer 2003 in Europe was due not only to scorching daytime temperatures but also to the fact that the nighttime temperatures had been rising nearly twice as fast as the daytime temperatures. The lingering nighttime warmth deprived people of normal relief from blistering daytime temperatures and the opportunity to recuperate from heat stress. A U-shaped relationship exists between temperature and mortality, resulting in more deaths at the extremes (Drake 2000). Air pollution linked to longer, warmer summers particularly affects those suffering from respiratory ailments, such as asthma, and cardiovascular problems. Temperature increases also contribute to an increase of ozone in the atmosphere. According to Paul Epstein and Christine Rogers (2004:6),

> Heat waves take a disproportionate toll on those living in poor housing lacking air conditioning, and those with inadequate social supports. The majority of those affected during the 1995 heat wave in Chicago, for example, were African-Americans living in substandard housing.

John Berger (2000:36–37) provides the following overview of the impact of climate change on the prevalence of certain diseases:

> Milder temperatures have contributed to the spread of mosquito-borne diseases in Africa. Richards Bay, South Africa,

for example, which was once malaria-free, had 22,000 cases in 1999. Malaria has also reached the highland areas of Kenya and Tanzania where it was previously unknown. In the Andes of Colombia, disease-carrying mosquitoes that once lived at altitudes no higher than 3,200 feet have now appeared at the 7,200-foot level.

While climate change is not the only factor involved, today it is estimated that there are 300 to 500 million cases of malaria each year in Africa, resulting in between 1.5 to 2.7 million deaths, more than 90 percent occurring among children under five years of age. Climate change appears to have contributed to the resurgence of various other epidemics, including cholera in Latin America in 1991, pneumonia plague in India in 1994, and the outbreak of hantavirus in the US Southwest in 1994. Tony McMichael (2001:302) presents the following sobering observations:

> The main anticipated impact of climate change on the potential transmission of vector-borne diseases would be in tropical areas. In general, populations on the margins of endemic areas in tropical and subtropical countries would be most likely to experience an increase in transmission. . . . This appears to reflect a combination of increasing population mobility, urbanization, poverty and regional warming, along with a slackening of mosquito control programmes. Meanwhile, in temperate zones, climate change may also affect diseases such as tick-borne viral encephalitis (which occurs in parts of Western Europe, Russia, and Scandinavia) and Lyme disease.

We can speak of the diseases of climate change or global warming. They include any "tropical disease" that spreads to new places and peoples, as well as failing nutrition and freshwater supplies because of desertification of pastoral areas or flooding of agricultural areas. The UN Food and Agriculture Organization has warned that in some 40 percent of the poorest developing societies with some 2 billion people, climate change may drastically increase the numbers of malnourished individuals. The growing tendency to power motor vehicles with biofuels, such as corn and sugarcane, which many see as emitting fewer greenhouse gas emissions, has already contributed to a global food crisis. In referring to developments in India, Vandana Shiva (2008:50) observes,

> Today, cars eat men [as well as women and children]. Land is diverted for parking, roads, highways, overpasses, and car factories. The mining of the iron ore and bauxite that makes the steel and aluminium is destroying the land and ecosystems. The atmosphere is being eaten up with fossil fuel emissions.

Conclusion

Climate change threatens to have serious impacts on many of the peoples whom anthropologists have historically studied, including foragers, horticulturalists, pastoralists, peasants, and more recently impoverished urbanites. Small indigenous and peasant communities in particular are finding themselves threatened by sea level rise, retreating glaciers, contraction of the Arctic ice cap, and loss of water supplies due to increased aridity or excessive precipitation, as well as diminished food supplies as native species are lost due to climate change. The loss of native species will have a drastic impact on dietary and nutritional patterns and thus on health. South Pacific Islanders, such as those living in Tuvalu and the Cataret Islands of Melanesia, particularly face a threat to their traditional horticultural lifestyle due to rising sea levels that inundate their fields and water supplies and threaten to submerge their islands. Rising temperatures are contributing to increased heat stress and the spread of infectious diseases to latitudes farther and farther north and south of the equator and to higher elevations. The effects of climate change, including the physical and mental health effects, such as the emotional disruptions associated with forced relocation from drought or flooded areas, will disproportionately impact poor nations and poor persons within all nations.

Once various feedback chains, related in part to the long lifetime of some greenhouse gases and the long memory of the climate chain, get started, they may be self-perpetuating and need no further anthropogenic input to keep going. While most projections of the impact of climate change only depict scenarios up until 2100, it is extremely important for both the sake of humanity and the planet to develop scenarios further into the future, to 2200, 2300, and beyond. After all, given that humanity has been around for some 5 to 6 million years, it is crucial that we have a long-term view of the impact of our ongoing existence on our fragile ecosystem, one that will ultimately, as I argue later in this book, require a vision of an alternative world system, one based on two cardinal principles—namely, social equity and justice and environmental sustainability. At the risk of sounding like a doomsday prophet, I maintain that it is evident that humanity needs to invoke the *precautionary principle,* which maintains that it is better to err on the side of caution than to allow the possibility of an even more serious catastrophe or a dystopian world than climate scientists have anticipated.

2

The Capitalist World System and Its Contradictions

Climate change constitutes one of the most important dilemmas of the twenty-first century, along with the growing gap between the rich and poor within and between nation-states, thanks to corporate globalization and ongoing conflicts in many parts of the world. This chapter explores in some detail the contradictions of the capitalist world system, including environmental degradation, of which climate change is the most profound form. Conventional proponents of capitalism laud its technological feats and proclaim that eventually technological innovations and economic growth will result in material prosperity for all. Conversely, anthropologist John Bodley (2008) contends that the global crisis provoked by corporate capitalism results in many social problems, including ongoing population growth, overconsumption, social stratification, environmental degradation, militarism, crime, and many personal crises, both physical and mental. Critical sociologist Erik Olin Wright (2010:36) succinctly and explicitly delineates 11 propositions critiquing capitalism as an economic system. Of these, Propositions 4, 6, 7, and 9 in particular are relevant to my discussion of the contradictions of the capitalist world system.

- Proposition 4: "Capitalism violates liberal egalitarian principles and social justice."

- Proposition 6: "Capitalism has a systematic bias towards consumerism."
- Proposition 7: "Capitalism is environmentally destructive."
- Proposition 9: "Capitalism, in a world of nation states, fuels militarism and imperialism."

Profit Making, Economic Growth, and the Treadmill of Production and Consumption

Capitalism is a global economic system that in its drive for profits requires ongoing accumulation and expansion. It systematically exploits human beings and the natural environment in pursuing its aims, despite rhetoric that it contributes to the prosperity and well-being of all human beings, albeit some more than others. Capitalism asserts that wealth generated at the top will eventually trickle down to the poor, thereby lifting them out of a tragic situation. Global capitalism fosters a treadmill of production and consumption primarily for the purpose of generating profits for a few and, in the process, because they are rated of lesser importance relative to profit making, sacrifices basic human needs and environmental sustainability. Machines of all sorts have played a central role in sustaining particularly industrial capitalism. Anthropologist Alf Hornborg (2001:2) contends that machine "power" entails "power to conduct work, power over other people, and power over our minds."

What Bodley (2008:95–98) terms the *culture of consumption* is an integral component of global capitalism, particularly in developed societies but also increasingly in many developing societies. As Don Slater (1997:121) aptly observes,

> Culture as a whole has become consumer culture. All culture is now produced, exchanged and consumed in the form of commodities. . . . All consumption . . . has become compensatory, integrative and functional. It offers the illusions of freedom, choice and pleasure in exchange for the real loss of these qualities through alienated labour; or integrates people within the

> general system of exploitation by encouraging them to define their identities, desires and interests them in terms of possessing commodities; and is functional in that consumer culture offers experiences ideally designed to reproduce workers in the form of alienated labour.

In order to survive, capitalism must generate an artificial need—namely, the need to endlessly consume a wide array of commodities, even potentially dangerous and lethal ones, such as motor vehicles, which emit pollutants and greenhouse gases. Sinclair Lewis, the renowned American socialist novelist, satirized emergent consumerism in *Babbitt*, an early twentieth-century novel in which he depicted a life of selling and consuming commodities as a one-way road to alienated mass conformity. Global capitalism and its associated ever-expanding cycle of production and consumption have fostered what Richard Hofrichter (2000) terms a *toxic culture*. According to Hofrichter (2000:1), "Elements of toxic culture might include the unquestioned production of hazardous substances, tolerance for economic blight, dangerous technologies, substandard housing, chronic stress, and exploitative working conditions." As we will see in the next chapter, the treadmill of production and consumption that results in many greenhouse gas emissions is contributing to the toxic culture that is an inherent component of global capitalism.

Robert Bocock (1993:2) argues that capitalism has held great appeal in the Western world and even the Soviet bloc because it seems to provide many people with a cornucopia of consumer items. Indeed, consumerism has provided an ideological rationale for capitalism. Under late capitalism, many people work not merely to subsist but in order to have the income to purchase an enormous array of consumer goods. As humanity has entered the twenty-first century, we have seen the enormous expansion of consumerism not only in the developed world but also in developing countries, such as China and India—by far the two most populous countries in the world. A. Fuat Firat and Nikhilesh Dholakia (1998:129) maintain that the vast majority of people in production jobs deem them "just a job" that they have to do in order to survive, thus prompting them to find compensation in the culture of consumption.

The Growing Socioeconomic Gap

Immanuel Wallerstein (1979:66) maintains that capitalism "as a system of production for sale in a market for profit and appropriation of this profit on the basis of individual or collective ownership has only existed in, and can be said to require, a world-system in which the political units are not coextensive with the boundaries of the market economy." Over the course of the development of global capitalism, the gap between developed and developing countries has tended to widen (UNDP 1999). The following lists the ratio of income between the richest fifth to the poorest fifth of countries for selected years:

- 1820 = 3:1
- 1870 = 7:1
- 1913 = 11:1
- 1960 = 30:1
- 1990 = 60:1
- 1997 = 74:1

The World Bank (2001), a capitalist institution that claims to be committed to the eradication of global poverty, reported that whereas in 1960 per capita GDP in the richest countries was 18 times greater than in the poorest countries, by 1995 this gap had widened to 37 times. The poorest 20 percent of the world's population earns only about 1 percent of the world's income (Ponting 2007:337). Hornborg asserts that under industrial capitalism, the "foundation of machine technology is not primarily about know-how but unequal exchange in the world system, which generates an increasing, global polarization of wealth and impoverishment." While there has been a tremendous amount of economic development in East Asia, South Asia, and Southeast Asia, much of it accompanied by widening social stratification and environmental degradation, Africa, despite being endowed with tremendous natural resources, has become home to the poorest people in the world. According to Maddison (2001), 57 African countries were worse off in terms of per capita GDP in 1998 compared with 1950. Table 2.1 shows the number of people living on less than US$2 per day.

The *Human Development Report* of 2005 stated that "during the 1990s, 25 countries in Sub-Saharan Africa and 10 in Latin America experienced a sustained period of economic stagnation" (quoted in Surin 2009:98). One scholar reports that whereas the top 20 percent of the world's population receives 75 percent of all income, the bottom 20 percent receives a mere 1.5 percent of all income (Taylor 2008:71). Invariably, wealth as opposed to income is always more concentrated, with the top 1 percent of the world's population owning 40 percent and the bottom 50 percent owning 1 percent of all wealth (Taylor 2008:1). Branko Milanovic (2002) reports that the richest 1 percent of people in the world receive as much income as the bottom 57 percent and that the top 10 percent of the US population earns an aggregate income that equals that of the world's population. Derek Wall (2010a:13) reports,

> Despite losing some of their wealth because of the recession, the world's three richest individuals—Bill Gates, Warren Buffet and Carlos Slim Helu—were worth $112 billion in 2009,

Table 2.1. People Living on Less Than US$2 per Day (in millions)

Region	*1981*	*2001*
South Asia	821	1,059
East Asia and Pacific	1,151	868
China	858	596
Sub-Saharan Africa	288	514
Latin America and Caribbean	99	128
Europe and Central Asia	8	93
Middle East and North Africa	52	70
Total	2,419	2,732
Total, excluding China	1,561	2,136

Source: Adapted from Anderson and Cavanaugh (2005:50).

> according to the *Forbes* list of the world's billionaires. *Forbes* recorded a total of 793 billionaires in 2009.

While the superwealthy people tend to be concentrated in developed countries, one also finds a growing number of the superwealthy in developing societies.

Growing socioeconomic inequality is characteristic of many societies around world, including developed countries. Table 2.2 compares the income inequality of selected developed countries.

Whereas in the United States, the gap between average CEO pay and worker pay was 45:1 in 1982, by 2003 it had grown to 301:1 (Anderson and Cavanaugh 2005:56). R. Wilkinson and K. Pickett (2009) reported in 2007 that CEOs of the "largest US companies received well over 500 times the pay of their average employees." This widening gap is the direct result of a shift of wealth from poor and working-class people to the rich. Holly Sklar (2006:A11) notes,

> In today's corporate America, workers see gutted paychecks and pensions despite rising worker productivity, while CEOs get golden pay, perks, pensions, and parachutes. The pay gap between average workers and CEOs has grown nine times wider since the 1970s. . . . The number of billionaires is at a

Table 2.2. International Comparison of Income Inequality

Country	*Richest 10 Percent to Poorest 10 Percent*	*Richest 20 Percent to Poorest 20 Percent*	*Gini Coefficient*
Japan	4.5	3.4	0.249
Sweden	6.2	4.0	0.250
Germany	6.9	4.3	0.283
France	9.1	5.6	0.327
Canada	9.4	5.8	0.352
Australia	12.5	7.0	0.352
United Kingdom	13.8	7.2	0.360
United States	15.9	8.4	0.408

Source: UNDP 2006: table 15.

> record high, but the share of national income going into wages and salaries is at a record low.

CEO compensation comes in various forms. Lee Raymond, the former head of ExxonMobil, for example, received a $398 million retirement package, pushing his earnings over a 13-year period to $686 million, or about $145,000 per day (Rumble 2006). The Global Financial Crisis or Global Recession has pushed millions of Americans out of the middle class, increased poverty, and resulted in many Americans losing their dwelling units.

Although not as socially stratified as the United States, Australia has become one of the most stratified developed societies in the world. According to Frank Stilwell (2000:81),

> Income inequality in Australia is relatively high by international standards, higher than Japan, for example, and most European countries with experience of social democratic governments. Wealth inequalities are even greater. The distribution of both income and wealth has become more unequal over the last two [now three] decades.

According to Meagher and Wilson (2008:227),

> In 1992, the average income of a top CEO was 27.2 times greater than the average annual money wage of an employed Australian. By 2002, the average remuneration of a top 50 CEO was 98.4 times the annual wage of an Australian worker.

The maldistribution of wealth is much more pronounced than that of income. The Australian Bureau of Statistics (2006:11) reports that wealthiest 20 percent of Australian households own 59 percent of total wealth, and the bottom 40 percent own a mere 7 percent of the wealth.

In many developing countries,

> many elites have gotten rich by snatching up state-owned enterprises at bargain rates when governments were forced . . . to privatize. In 2003, Asia (excluding Japan) and Latin America boasted seventy-six of the world's 587 billionaires, up from

> only five in 1986. This doesn't even include Russia's billionaires, who were nonexistent in 1986 but numbered twenty-four as of 2003. (Anderson and Cavanaugh 2005:56)

Despite its alleged commitment to social parity as a supposed socialist market economy, since the late 1970s China under modernization has experienced a marked increase in social inequality:

> Whereas inequality in China in 1980 was comparable to that of social democratic Germany (Gini coefficient = 0.25), by 2005 it was less equal than Russia (Gini coefficient = 0.45). The wealthiest 10 percent of the Chinese population earned seven times that of the poorest 10 percent in the 1980s but by 2005 that inequality had risen to a factor of 18. The richest 10 percent of the population now accounts for 45 percent of the country's wealth, the poorest only 1.4 percent. (Smith 2010:256)

Depletion of Natural Resources and Environmental Degradation

To a greater or lesser degree, all human societies encroach on and modify the natural environment. Foragers contributed to the creation of grasslands, pastoralists overgrazed their lands, and peasants caused deforestation. Jared Diamond (2005) maintains that environmental degradation was one of the principal factors contributing to the demise of the Anasazi in the US Southwest, the chiefdom societies of Easter Island, and the Maya civilization of the Yucatan Peninsula and Guatemala. The emergence of social mechanisms for harnessing large amounts of energy from the environment contributed to what Eugene Ruyle (1977:623) terms *predatory ruling classes*. The dangers of ecological self-destruction that plagued ancient societies became even more pronounced with the advent of industrial capitalism around 1800 and has grown ever greater with the passage of time. The Industrial Revolution allowed for the harnessing of nonrenewable fossil fuels—namely, coal, petroleum, and natural gas—and humans came to rely increasingly on machine power rather than the energy derived from humans, animals, and even water and wind.

Rosa Luxemburg ([1913] 1959) in *The Accumulation of Capital* argued that capital accumulation would lead to environmental degradation. This has been recognized even by non-Marxian scholars. Forty years ago, Donnella Meadows et al. (1972), in their seminal book *The Limits to Growth*, recognized that capital accumulation would prove environmentally unsustainable by 2100 if unchecked. A 30-year update for the book essentially confirms the original findings (Meadows, Randers, and Meadows 2005). The *Limits to Growth* project was part of a larger study called the Project on the Predicament of Mankind sponsored by the Club of Rome, a collection of prominent business executives and mainstream economists from the United States, Western Europe, and Japan who were assembled by Fiat executive Aurelio Peccei. The *Limits to Growth* project advocated a no-growth economy and a redistribution of income sufficient to guarantee everyone a subsistence income while maintaining a capitalist economic system, a stance that contradicts the need for capitalism to grow or die out. According to Resistance (1999:43), an Australian-based socialist group,

> Under capitalism, no-growth periods are produced by the inherent tendency of capitalist production to outstrip the market. During such crises (commonly called recessions) the owners of industry show no inclination to distribute the necessities of life to workers thrown on the scrapheap by production cutbacks.

In essence, capitalism lacks a concept of sufficiency, of individuals having enough, and demands continual growth. If it does not grow, it may implode in an economic crisis, as occurred during the Great Depression of the 1930s.

Three decades ago Andre Gorz (1980:11–12) argued that capitalism is on the verge of self-destruction because of its emphasis on ever-expanding production:

> Economic growth, which was supposed to ensure the affluence and well-being of everyone, has created needs more

> quickly than it could satisfy them, and has led to a series of dead ends which are not solely economic in character: capitalist growth is in crisis not only because it is capitalist but also because it is encountering physical limits. . . . It is a crisis in the character of work: a crisis in our relations with nature, with our bodies, with future generations, with history: a crisis of urban life, of habitat, of medical practice, of education, of science.

Table 2.3 depicts the annual energy consumption for various world regions in 2004.

Danny Chivers (2009:103) reports the following approximate figures of energy use per capita (kWh) in selected countries in 2008: Canada, 96,000; United States, 89,000; Australia, 75,000; European Union, 48,000; China, 19,000; Tanzania, 4,000; and Nepal, 3,500. Bear in mind, in all of these countries, there are class differences in terms of energy utilization. Michael Klare (2008:33) observes that the "worldwide requirement for primary energy is expected to rise by 57 percent between 2004 and 2030," with much of the growth occurring in Asia, particularly China and India.

Table 2.4 shows the ecological footprint of various categories of countries, regions, and selected specific countries. The term *ecological footprint* refers to a measurement in global hectares that aggregates six types of productive areas: (1) cropland, (2) graz-

Table 2.3. Annual Energy Consumption for Various World Regions, 2004

Region	*Percentage Share of World Energy Consumption*
Asia and Oceania	31
North America	27
Europe	19
Eurasia	10
Middle East	5
Central and South America	5
Africa	3

Source: Adapted from Burman (2007:20).

ing land, (3) forest, (4) fishing ground, (5) built-up land, and (6) land area required to absorb CO_2 for fossil fuel use (Wackernagel and Rees 1996).

The United States has the highest per capita ecological footprint of any country, considerably higher than many developed societies, particularly in Europe. Specific countries vary widely in the ecological footprints of specific individuals. In a developed country, a wealthy family with multiple residences and motor vehicles and frequent holidays in far-off places leaves a much greater ecological footprint than a slum dweller or homeless person does. A profound form of environmental degradation around the world is deforestation. Deforestation was once primarily a phenomenon of temperate regions, such as Europe and North America, but occurs now primarily in the tropical zones of the developing countries, such as Brazil, Indonesia, and Southeast Asian countries. The 3.6 billion hectares (a hectare is 2.41 acres) of forest in existence in 1980 declined by 5 percent to 3.2 billion hectares by 1995 (Rosa and Dietz 2010:23). Urban sprawl has been a major factor accounting for deforestation.

Poverty, Population Growth, and the Environment

The United Nations (2005) projects an increase in population by 2050 of somewhere between 7.68 billion as the low variant and 10.65 billion as the high variant, with 9.08 billion as the medium variant. Demographers and other scholars have often observed that the greatest rate of population growth tends to occur among the poor. According to Dudley Poston Jr. and Leon Bouvier (2010:368),

> An ever-growing proportion of [the global] population resides in the developing regions of the world. More than 95 percent of the projected growth in the world population between 2010 and 2050 is expected to occur in developing countries. By 2050, 85 percent of the world's projected population will hail from developing countries. This population growth will come

Table 2.4. Ecological Footprint of World Regions and Selected Countries, 2007

	Population (million)	*Ecological Footprint (global hectares per capita)*
World	6,671.6	2.7
High-income countries	1,031.4	6.1
Middle-income countries	4,323.4	2.0
Low-income countries	1,303.3	1.2
Unclassified countries		13.5
Africa	**963.9**	**1.4**
Congo, Democratic Republic of	62.5	0.8
Egypt	80.1	1.7
Ethiopia	78.6	1.1
Kenya	37.8	1.1
Nigeria	147.7	1.4
South Africa	49.2	2.3
Asia	**4,031.2**	**1.8**
Afghanistan	26.3	0.6
Bangladesh	157.8	0.6
China	1,336.6	2.2
India	1,164.7	0.9
Indonesia	224.7	1.2
Japan	127.4	4.7
Kuwait	2.9	6.3
Nepal	28.3	3.6
Pakistan	173.2	0.8
Qatar	1.1	10.5
Timor-Leste	1.1	0.4
United Arab Emirates	6.2	10.7
Europe	**730.9**	**4.7**
Albania	3.1	1.9
Denmark	5.4	8.3
Germany	82.3	5.1
Russian Federation	141.9	4.4
Spain	44.1	5.4
United Kingdom	61.1	4.9
Latin America and Caribbean	**569.5**	**2.6**
Bolivia	9.5	2.6
Brazil	190.1	2.9
Cuba	11.2	1.9
Haiti	9.7	0.7
Mexico	107.5	3.0
Peru	28.5	1.5
Uruguay	3.3	5.1
Venezuela	27.7	2.9
North America	**341.6**	**7.9**
Canada	32.9	7.0
United States	308.7	8.0
Oceania	**34.5**	**5.4**
Australia	20.9	6.8
New Zealand	4.2	4.9
Papua New Guinea	6.4	2.1

Source: Adapted from Global Footprint Network 2010.

> disproportionately from people on the margins, those with limited resources and life opportunities.

While upper- and middle-class people worldwide frequently are perplexed as to why the poor have more children—because the well-off realize that financially supporting children to adulthood and often beyond constitutes a massive expenditure—poor people in developing countries, both in rural and urban areas, commonly view children as breadwinners who add to the family coffers. As none other than Jeffrey Sachs (2008:16) observes, population

> will soar in precisely those parts of the world that are struggling the most today with extreme poverty, disease, famine, and violence. Both cause and effect are at play. Poverty contributes to high fertility rates, while high fertility rates prolong poverty. The poorest countries in the world are stuck in a demographic trap as much as a poverty trap.

Mahomood Mamdani (1972) maintains that peasants in India and other developing societies particularly welcome additional sons because they can help work the land. Even in urban areas, children are viewed as an asset among the poor. Boys in Mexican cities, for example, often work as shoe shiners, and both young boys and girls sell items such as chewing gum, tortillas, and other foods prepared by their mothers. The reality of how young children in developing countries function as breadwinners was vividly brought home for me in the summer of 1982 during a weeklong stay in Antigua, the former colonial capital of Guatemala. While there, I purchased "La Familia"—a set of dolls consisting of a peasant family made up of father, mother, son, and daughter—from a girl of around five for the equivalent of US$1. Antigua, a major tourist destination and center of Spanish-language schools, was particularly hard hit economically at the time because of a civil war overseen by the Pentecostal dictator Rios Montt, who enjoyed US backing.

Historically, capitalist penetration of indigenous societies and precapitalist state societies contributed to population increases, as local households struggled to meet the demand for taxes imposed by colonial powers. In this context, having more laborers and wage earners became an important household strategy for survival in a globalizing capitalist system. Rural household financial difficulties often pushed males, sometime females, and even entire families to migrate to urban areas to find employment. As Ian Angus and Simon Butler (2011:212) observe, "High birth rates aren't the cause of third world poverty—they are an effect of poverty, and building birth clinics, however important that is for other reasons, won't eliminate the underlying causes." Containing population growth ultimately will entail creating a much more equal playing field in terms of access to resources of all sorts.

Resource Wars

While states and empires have long engaged in "resource wars," as Michael Klare so aptly observes in *Resource Wars* (2001), the discovery of oil in the late nineteenth century added a new dimension to warfare. According to Gabriel Kolko (2006:177),

> The destructive potential of weaponry has increased exponentially, and many more people and nations have access to it. . . . The world has reached the most dangerous point in recent, or perhaps all of, history. There are threats of war and instability unlike anything that prevailed when a Soviet-led bloc existed.

Since the beginning of the twentieth century, oil has played a part in both world and regional wars. Despite the religious and political rationales often used as ideological justifications for war, violent conflict between societies, both prestate and state, have often entailed the struggle for economic resources. In terms of the rich oil reserves of the Persian Gulf region, as Klare (2004:x) observes,

> first to spar were Great Britain and czarist Russia, later joined by France, Germany, and the United States. By the end of the twentieth century, safeguarding the flow of oil from the Persian Gulf had become one of the most important functions of the U.S. military establishment.

The Japanese attack on Pearl Harbor on December 7, 1941, was in part prompted by the American decision to cut off oil exports to Japan earlier that year (Heinberg 2006:54). Before this event, Japan had relied very heavily on imported oil from the United States and invaded the Dutch East Indies in part to access its rich oil fields. The 1953 coup in Iran orchestrated by the United States and United Kingdom, resulting in the overthrow of Prime Minister Mohammad Mossadegh, was in large part prompted by his call to nationalize the country's oil fields.

Paul Baran and Paul Sweezy (1966:183) argue that the United States had employed its military and economic power to "attract large segments of old colonial empires into its own neo-colonial empire." In its assertion that it formed the principal bulwark against the spread of communism, the United States, in the wake of World War, (1) provided economic support to various capitalist powers, particularly the United Kingdom, West Germany, and Japan; (2) created an elaborate system of military alliances and bases around the massive perimeter of the Soviet bloc; and (3) developed a massive military force that essentially functioned as a form of state capitalism under the guise of what Dwight Eisenhower aptly termed the *military-industrial complex* (Baran and Sweezy 1966:191). In contrast, while the Soviet Union itself developed its own military-industrial complex, its stance in foreign relations was largely defensive, even to the point that it used its military muscle to create a buffer zone between itself and the West by establishing and maintaining satellites in Eastern Europe.

Tensions in the Middle East constitute the single most significant factor affecting the price of oil over the course of the past three decades or so. As the editors of *Monthly Review* (2002:9) reported a few months before the US invasion of Iraq in 2003,

> Military, political, and economic aspects are intertwined in all stages of imperialism, as well as capitalism in general. However, oil is the single most important strategic factor governing U.S. ambitions in the Middle East.

Middle Eastern governments, such as Saudi Arabia, Iran, and Iraq, have used their oil earnings to purchase more weapons and have helped arms manufacturers in capitalist developed societies earn tremendous profits. What Nitzan and Bichler (2002:198–263) dub the "Weapondollar-Petrodollar Coalition" requires an "atmosphere of permanent threat," in much the same way that the Soviet Union did during the Cold War. At the same time, while the United States has acted in an increasingly imperialist manner in the Middle East, the war in Iraq has split the countries of the European Union in various ways and revealed internal tensions within the core countries of the capitalist world system (Boswell 2004:523).

Furthermore, China, with its concerted state capitalist program of industrialization, modernization, and economic expansion, and Russia also constitute major powers with a strategic interest in access to petroleum, an interest shared by other major powers, such as Germany, the United Kingdom, France, and Japan, as well as virtually all minor powers in both the developed and developing worlds. The Caspian Sea Basin, with an estimated fifth of the world's total proven oil reserves, constitutes yet another potential hot spot in the struggle for a diminishing natural resource. Darfur in southern Sudan and other parts of sub-Saharan Africa constitute the foci for a new cold war, one focusing on competition for oil between Western powers and China, which requires massive amounts of this commodity to support is monumental economic growth (Engdahl 2007). Elsewhere in sub-Saharan Africa, the United States is now obtaining more than 15 percent of its oil from various West African countries, almost as much as it obtains from Saudi Arabia. While US oil companies were signing contracts with Gabon, Nigeria, Equatorial Guinea, Angola, and Algeria, the George W. Bush administration increased aid to and stationed military advisors in various African countries (Turshen 2004:2).

Another cold war of sorts has developed between the United States and various South American countries, particularly Venezuela, over the past several years. As part of his anti-imperialist policies, Venezuelan president and populist Hugo Chávez has used the oil income of his country's nationalized oil industry to finance ambitious social programs, especially in education and health care. He has agreed to provide Cuba with 53,000 barrels of oil a day (Jones 2007:288). Financially strapped Cuba will be allowed to pay with a combination of money, goods, and services. This arrangement is estimated to be worth $550 million a year and will supply Cuba with one-third of its needed oil supply. Venezuela has also struck an agreement in which Venezuela will sell China oil and obtain 18 ships from China (Jones 2007:443). Under Chávez's presidency, Venezuela has encouraged its hemispheric neighbors to create regional oil consortia for the Caribbean (Petrocaribe), the Andean region (Petroandino), South America (Petrosur), and Latin America (Petroamerica).

Conclusion

Bearing in mind the gravity of climate change, any effort to grapple with its anthropogenic sources must recognize the contradictions of global capitalism. Over the course of the development of global capitalism, the gap between rich and poor countries in terms of access to income and wealth has tended to widen, as has the gap between the rich and poor within most nation-states. Despite the end of the Cold War in the wake of the collapse of the Soviet bloc, conflicts have occurred around the world, which in part can be related to various states, led by the United States but including the United Kingdom and Australia, that are willing to do the bidding of multinational corporations. Capitalism, with its emphasis on economic expansion and ongoing production, operates on the assumption that natural resources are infinite, when in reality many of them are finite. It contributes to depletion of natural resources and environmental degradation, including climate change or global warming. As an economic system of unequal exchange, global capitalism results in overdevelopment for some and underdevelopment for others. Capitalism is a system in which wealth and poverty are intricately related. In their quest for survival, many poor people around the world bear high numbers of offspring, which also strains the environment, but not nearly to the same extent as the high levels of consumption among the affluent sectors of the global economy. Multinational corporations and state companies in both capitalist and postrevolutionary societies have created not only a global factory but a new global ecosystem characterized by industrial and motor vehicle pollution, toxic and radioactive wastes, deforestation, desertification, and last, but not least, climate change. As a result of its emphasis on ever-expanding production, global capitalism is on the verge of self-destruction and in the process of contributing to the destruction of much of humanity and a fragile ecosystem.

3

The Capitalist Treadmill of Production and Consumption as a Generator of Greenhouse Gas Emissions

Climate change, perhaps more than any other environmental crisis, illustrates the contradictions and unsustainability of global capitalism. According to Bodley (2008:307),

> One crucial growth trend is for scale increases in a country's per capita GDP to correlate strongly with scale increases in carbon dioxide emissions per capita. . . . This correlation points to a link between economic growth and global warming and environmental degradation.

Developed countries have tended to shift toward service/information economies and have the financial resources to invest in more energy-efficient technologies, which can over the long run cut costs. Multinational corporations in the core often export or outsource polluting industries to developing countries.

Table 3.1 indicates that while the total CO_2 emissions in high-income countries greatly exceeds that in the high-middle-, middle-, and low-income countries combined, the production efficiency in all of the latter countries is worse than that for the high-income countries. As J. Timmons Roberts, Peter Grimes, and Jodie Manale (2003:288) observe, developing countries have "enough fossil-fuel dependent technology to compete in the world market, but not enough sophisticated infrastructure to do so efficiently."

Table 3.1. Average Total CO_2 Emissions and CO_2/Unit GDP for Income Groups of Countries, 2000

Category of Countries	*Average Total Emissions*	*Average Cumulative CO_2 (million tons CO_2)*	*Unit GDP*
High income	24	120,162	0.1479
High-middle income	20	5,917	0.2710
Middle income	33	18,161	0.2960
Low income	54	20,155	0.5262
Low income	61	28,834	0.4066

Source: Adapted from Roberts and Parks (2007:147).

While the production efficiency has tended to improve in developed countries, there has also been a tendency for total CO_2 emissions and per capita emissions to increase, as occurred in the United States, the Netherlands, Japan, and Austria between 1975 and 1996 (Clark and York 2005:412). Such a trend is consistent with the need of global capitalism to grow continually. Roberts, Grimes, and Manale (2003:277–78) identify various social factors underlying CO_2 intensity of production within countries, as defined by quantity of CO_2 released per unit of economic output. They found that some countries had been more efficient than others in generating wealth for the environmental cost entailed. Elsewhere, Bert Metz (2010:39) reports that Japan and European countries have tended to be the most efficient economies, being about 25 percent more efficient than the United States and more than three times as efficient as economies in transition (Metz 2010:39). While the West German economy grew at an average annual rate of 2.1 percent during the 1980s, its CO_2 emissions declined an average of 1.2 percent per annum. Advanced capitalist or developed countries are both economically able and politically pressured by the environmental movement to reduce pollution (Roberts, Grimes, and Manale 2003:285). In 2005 the top 10 emitters of greenhouse gases (GHGs) were either specific developed countries (such as the United States, Japan, and Canada) and the European Union or the "large emerging market economies" (such as China, Russia, India, Brazil, Mexico, and

Indonesia) (Barbier 2010:35). Altogether they accounted for over 70 percent of the world's greenhouse gas emissions.

William Stanley Jevons identified an important paradox—namely, that increasing efficiency of coal use was correlated with increasing coal consumption (York, Rosa, and Dietz 2009:137). Paradoxically, the most "eco-efficient" businesses, industries, or economies may be the ones consuming the largest quantities of resources and producing the most pollution. For example, while China has undergone a significant improvement in energy efficiency, it has undergone a marked increase in its total ecological footprint (York, Rosa, and Dietz 2009:140). These trends challenge the basic assumptions of ecological modernization theory, which maintains that technological transformations will be sufficient in solving environmental problems. Four countries declined in ecological footprint intensity between 1961 and 2003 but increased in terms of total ecological footprint (York, Rosa, and Dietz 2009:142).

Table 3.2 presents CO_2 emissions from various broad sources, and table 3.3 shows those from more specific sources. In terms of historical responsibility for global carbon emissions between 1750 and 2006, the United States accounted for 28 percent; the United Kingdom, 6 percent; Japan, 4 percent; Russia, 8 percent; Germany, 7 percent; the remainder of Europe, 18 percent; and

Table 3.2. Total Carbon Dioxide Emissions (in Gigatons and Percentage of Total), 2007

Source	*Gigatons*	*Percentage*
Fossil fuels	29	81
Electricity	11.5	32
Industry	8	22
Transportation	6.5	18
Residential	2	6
Commercial	1	3
Deforestation	7	19
Total	36	100

Source: Adapted from Sachs (2008:96). These figures are based on fossil fuel estimates for 2005 compiled by the International Energy Association in 2007 and assume that all categories increased by 2.3 per annum between 2005 and 2007.

Table 3.3. Approximate Breakdown of Humanity's Global Greenhouse Gas Emissions, 2009

CO_2 Source	*Billion Tons*	*Percentage CO_2e per Year*
Coal—electricity and heating	8.0	17.3
Coal—industrial use	4.6	9.8
Oil—overland transport	6.0	12.9
Oil—shipping	1.2	2.6
Oil—aviation	0.9	2.0
Oil—other	1.7	3.6
Oil—industrial use	1.2	2.6
Gas—electricity generation	1.8	3.8
Gas—heating	1.9	4.1
Gas—other	0.2	0.4
Other fuels	0.1	0.3
Cement manufacture CO_2 (nonenergy)	1.9	4.1
Land-use change CO_2	4.4	9.4
N_2O from fertilizer use	2.4	5.2
Methane and N_2O from livestock	2.5	5.4
Methane from rice paddies	0.7	1.5
Methane and N_2O from other agriculture	0.8	1.7
Other greenhouse gases	0.5	1.0
Methane and N_2O from fossil fuel use	2.3	5.0
Total	**46.6**	**100.0**
Of which CO_2	35.7	77
Of which other greenhouse gases	10.9	23

Source: Adapted from Chivers (2009:73).

China, 8 percent (Schor 2010:175). Conversely, "per resident, the UK, the US, Russia, Belgium and Germany [in that order] have the largest historical responsibility for CO_2 emissions, at around 1,000 tonnes per person living today," in contrast to India and China, which "don't even make it into the top 20" (Chivers 2009:89).

A September 2008 report indicated that "2008 promised to have the highest rate ever" of CO_2 entering the atmosphere, despite the economic downturn of 2007 and 2008, along with efforts to lower emissions (Ward 2008:63). According to Ward (2008:63), the "published figures for 2007 showed a 3 percent increase in the amount of carbon put into the atmosphere compared to the year before." The Global Carbon Project (2010:1) reports that "the annual growth rate of atmospheric CO_2 was 1.6 ppm in 2009, below the average for the period 2000–2009 of 1.9 ppm per year (ppm = parts per million). The mean growth rate for the previous 20 years was about 1.5 ppm per year."

While the Global Financial Crisis played a role in bringing down the annual growth rate, the tail of La Niña appears to have resulted in increased land and ocean CO_2 sinks. CO_2 emissions in various developed countries declined appreciatively in 2009: United States, 6.9 percent; United Kingdom, 8.6 percent; Germany, 7 percent; Japan, 11.8 percent; and Russia, 8.4 percent (Global Carbon Project 2010:3). The Global Carbon Project (2011: 3) reported that there was a 5.9 percent increase in CO_2 emissions during 2010.

Table 3.4 indicates that CO_2 emissions from fuel combustion overall have continued to rise around the world, although they have dropped in specific regions, particularly former Soviet bloc countries or "economies in transition," and specific countries. Particularly striking is the tremendous increase in CO_2 emissions in both China and much of the rest of Asia, including India.

As table 3.5 indicates, when cement production is included into CO_2 emissions, China substantially outstrips the United

Table 3.4. World CO_2 Emissions in Gigatons from Fuel Combustion, Selected Countries and Regions

	1990	*2008*	*Percentage Change, 1990–2008*
United States	4,868.7	5,923.6	15.7
Europe	3,153.6	3,222.9	2.2
Pacific	1,346.4	1,582.0	17.5
Economies in transition	3,852.9	2,624.3	–31.9
Africa	545.6	889.9	61.1
Middle East	592.5	1,492.3	151.8
Non-OECD Europe	106.1	92.2	–13.1
Latin America	869.5	1,476.5	69.8
Asia (excluding China)	1,510.1	3,524.1	133.4
China	2,244.4	6,550.5	191.9
World	20,964.8	29,381.4	40.1

Source: International Energy Agency (2010:13).

Table 3.5. Total and Per Capita Annual CO_2 Emissions (from Fossil Fuels and Cement Production) of the 20 Highest-Emitting Countries, 2009

Country	*Annual Emissions (million tons of CO_2)*	*Percentage of Global CO_2 Emissions*	*Tons of CO_2 per Capita*
China	8,060	26.0	6.1
United States	5,310	17.1	17.2
India	1,670	5.4	1.4
Russia	1,570	5.1	11.2
Japan	1,180	3.8	9.2
Germany	770	2.5	9.3
South Korea	560	1.8	7.7
Canada	540	1.7	16.3
United Kingdom	490	1.6	8.1
Mexico	470	1.5	4.2
Indonesia	440	1.4	1.9
Italy	410	1.3	7.0
Australia	400	1.3	18.8
Brazil	380	1.2	1.9
South Africa	380	1.2	1.9
Saudi Arabia	370	1.2	13.6
France	370	1.2	6.0
Spain	310	1.0	8.0
Total	**24,560**	**79.2**	

Source: Adapted from Chivers (2009:84).

States. Conversely, on a per capita basis, the United States emits 2.82 times more emissions than China. Australia exceeds the United States in terms of CO_2 emissions per capita in large part because of its very high reliance on coal-fired plants for electricity. Australia leads all developed countries in terms of deriving most of its electricity generation, 91 percent, from fossil fuels. In contrast, the United Kingdom derives 78 percent from fossil fuels; the United States, 71 percent; Germany, 61 percent; Japan, 51 percent; Austria, 33 percent; Canada, 25 percent; France, 10 percent; Switzerland, 2 percent; and New Zealand, 1 percent (Tiffen and Gittins 2009:158).

Table 3.6 indicates that various developing countries, particularly ones in the Middle East that produce high amounts of oil and natural gas, actually outstrip the leading developed countries in terms of CO_2 emissions per capita. The reason for this

Table 3.6. Ranking of World's Top 12 CO_2 Emitters in Metric Tons of Carbon per Capita, 2007

Rank	*Country*	*CO_2 per Capita*
1	Qatar	14.02
2	Kuwait	9.30
3	Netherland Antilles	8.79
4	United Arab Emirates	8.44
5	Bahrain	8.06
6	Trinidad and Tobago	7.58
7	Aruba	6.29
8	Luxembourg	6.16
9	Brunei	5.32
10	Falkland Islands	5.25
11	United States	5.20
12	Australia	4.84

Source: Tom Boden, Gregg Marland, and Bob Andres of the Carbon Dioxide Information Analysis Center, Oak Ridge National Laboratory (http:cdiac.ornl.gov).

will become apparent later in this chapter when I examine the ecological footprint of the United Arab Emirates. In this table, in contrast to table 3.5, the United States has a slightly higher CO_2 emissions per capita figure than Australia, perhaps due to differences in the calculations of the rates in the two tables.

Much of the economic growth in developing countries, including China and India as well as the oil-producing countries, results from the production of luxury goods for the wealthy and "new middle class" in those places. In China, many of these people can be found in the eastern provinces, where the individual purchasing power exceeds US$7,000 annually (Harris 2010:126). The Netherlands Environment Assessment Agency calculated that in 2006 China had surpassed the United States in total CO_2 emissions (Camilleri and Falk 2010:279). China builds a new coal power plant approximately every four days and appears to have surpassed the United States to become the single largest emitter of CO_2 in the world due to rapid growth in fossil fuel consumption and cement manufacturing.

Roberts and Parks (2007) delineate four factors contributing to differences in national responsibility for climate change: (1)

national wealth (per capita GDP or total GDP); (2) CO_2 emissions per unit of GDP (carbon intensity); (3) per capita CO_2 emissions; and (4) cumulative emissions over 50 years starting in 1950. Various studies indicate a curvilinear relationship in which intensity first rises as one moves from poorer to wealthier nations, but then declines among the very wealthiest nations. International trade and globalization shunt the greatest ecological impacts of production from developed societies to developing societies (Roberts and Parks 2007:161). In terms of historical responsibility, China and India combined have contributed only 9 percent of the total quantity of accumulated anthropogenic green house gas emissions, whereas the United States has contributed over 30 percent.

The Heavy Reliance of Global Capitalism on Fossil Fuels

Global capitalism is heavily reliant on fossil fuels as table 3.7 indicates. Indeed, in the 1930s Lewis Mumford (1934) described the world system as a form of "carboniferous" capitalism. The International Energy Agency (2010) projects that at present rates of growth, global energy use will rise more than 50 percent by 2030, with fossil fuels remaining the primary energy source. As environmental anthropologist Emilio Moran (2010:5) astutely observes, "The exploitation of the huge amounts of fossil fuel materials stowed away for geologic periods of time in sub-terrestrial sinks and the launching of the by-products from their use into the biosphere kicked off biogeochemical changes in the atmosphere that took a couple centuries to be felt and which now threaten our planet."

According to Andrew McKillop (2005:198), "We have near-total dependence on oil and natural gas for food, plastics, fertilizers, pharmaceuticals, mining the metals for our transport equipment, and running our city skyscrapers—in fact for everything we call 'advanced industrial.'" While China has increased its consumption of gasoline or petrol tremendously due to a

Table 3.7. World Fossil Fuel Consumption, 1950–2005

Year	*Oil*	*Natural Gas*	*Coal*
	(million tons of oil equivalent)		
1950	470	171	1,074
1955	694	266	1,270
1960	951	416	1,544
1965	1,530	632	1,486
1970	2,254	924	1,553
1975	2,678	1,075	1,613
1980	2,972	1,304	1,814
1985	2,801	1,493	2,107
1990	3,136	1,774	2,270
1995	3,252	1,938	2,282
2000	3,537	2,192	2,381
2005	3,859	2,475	2,930

Source: Adapted from Sawin and Mukherjee (2008:33).

tremendous growth in the number of automobiles, diesel fuel "doubled between 1997 and 2007, reaching levels three to four times that of gasoline" (Montgomery 2010:74). Three factors contributed to this increase in diesel fuel utilization: (1) a rapid increase in the number of trucks, partly due to the inability of existing rail systems to accommodate the increase in the volume of freight; (2) a growth in construction equipment needed to build new buildings, highways, and other forms of infrastructure; and (3) the "widespread use of diesel generators, for irrigation pumps and local electricity, due to an insufficient grid system" (Montgomery 2010:75).

Coal

The Chinese reportedly were using coal as early as 1000 BC and were using it for smelting iron since at least the fourth century (White 2008:107). The Romans burned it in Britain, but coal production began to take off in Europe when Henry III in England granted Newcastle free laborers a grant to mine it. Coal was used in various industries, including lime burning, brick making, brewing, soap making, glass making, and distilling.

Newcastle eventually emerged as an early center of the coal industry from whence coal was shipped to London, other English cities and towns, and eventually other parts of Europe. In time, capitalist and postrevolutionary societies came to rely heavily on coal. In Europe and North America, coal production skyrocketed during the second half of the nineteenth century. For example,

> in France between 1850 and 1873 the production or use of coal leapt from 7 million tons to nearly 25 million. In Germany it went from 5 million to 36 million. In the United Kingdom, the dominant power of the day, it soared from 37 million tons to 112 million. In Belgium, too, it tripled. (Simms 2009:97)

Coal in large part replaced charcoal derived from trees. In terms of iron production, whereas in 1825 France used 194 tons of charcoal as opposed to 5 tons of coke, a coal derivative, by 1885, France was using 1,600 tons of coke as opposed to 29 of charcoal (Simms 2009:97). While the United Kingdom was clearly the dominant coal producer worldwide in 1860, it had been superseded by the United States in this regard by 1910. In that year, in terms of millions of metric tons, the United States produced 517.0 compared with 292.0 for the United Kingdom, 277.3 for Germany, 40.1 for France, and 22.8 for Belgium (White 2008:109). Table 3.8 depicts the astronomical increase in coal production over the course of a century between 1860 and 1960.

The heavy reliance on coal was due in part to the fact that railways initially relied upon coal-fired steam locomotives,

Table 3.8. World Coal Production, 1860–1960

Year	*Million Tons of Coal*
1860	132
1880	314
1900	701
1920	1,193
1940	1,363
1960	1,809

Source: Adapted from Behringer (2010:175).

which have been a source of both fascination and perhaps horror in that they belched volumes of steam and smoke as they ripped through the countryside of Europe, North America, and later Asia, Africa, South America, and Australia. As Edwin Black (2006:19) observes, "For mankind, coal was both a blessing and a curse. This black combustible was plentiful in the ground. It was also poison in the air." Not only does coal contribute to air and water pollution, but it is a major source of carbon dioxide and contributor to climate change.

British Petroleum (BP), which now calls itself "Beyond Petroleum," estimates that the "world possesses proven reserves of 909 billion metric tons of coal, about half in the form of anthracite and bituminous coal, the other half in less desirable subbitumous coal and lignite" (Klare 2008:50). The leading reserves of coal are situated in the United States (246.6 billion metric tons), Russia (157.0 billion), China (114.5 billion), India (92.4 billion), and Australia (78.5 billion). The leading consumers of coal are China (38.6 percent), the United States (18.4 percent), India (7.7 percent), Japan (3.9 percent), and Russia (3.6 percent) (Klare 2008:51). The Global Carbon Project (2010:2) reports that coal constitutes the largest source of fossil fuel CO_2 emissions, with about 92 percent of the growth in coal emissions from 2007 to 2009 coming from increased coal use in China and India. Between 2002 and 2009, consumption of coal in China increased at an average rate of over 9 percent per annum (Montgomery 2010:103). Two-thirds of China's electricity comes from coal (Leonard 2008:40). China is building two coal-fired power plants every week (Li 2008:64). India is also quickly adding coal-fired power plants.

Oil

In *Modern Capitalist Culture*, anthropologist Leslie A. White (2008:118) observes,

> Coal might well be called the king, or the father, of the Fuel Revolution but two other fossil fuels have played important roles in this drama, also: petroleum and natural gas. As a

> matter of fact, petroleum has increased greatly in relative importance during the last century [the twentieth century] while coal has decreased relatively in significance.

Although the discovery of oil, which occurred in various places during the nineteenth century, preceded the advent of the automobile era by about half a decade, the rises of the petroleum and automobile industries were intricately interwoven. The invention of the internal combustion engine by Nikolaus Otto (1832–1891) paved the way for the invention of the motor car by Gottlieb Daimler and Carl Benz in the late 1880s (Behringer 2010:176). Oil also came to fuel diesel engines used in train locomotives, trucks, ships, airplanes, and even some cars. Aside from its use in transportation, oil has been the principal source of economic growth during most of the twentieth century and continues to be so in the twenty-first century (Heinberg 2006:1). Indeed, oil replaced coal as the primary fossil fuel in the 1950s. Oil also became an essential ingredient in the production of plastics, which in large part replaced packing materials such as paper, wood, glass, and metal. Oil in many instances replaced coal in the heating of dwelling units, offices, and factories. Table 3.9 illustrates the world's leading oil producers in 2005 and the projected leading oil producers in 2030 (see Klare 2008:42).

In terms of the next top 10 oil producers, Kuwait produces 3.3 percent; Algeria, 2.6 percent; Canada, 2.4 percent; Iraq, 2.3 percent; United Kingdom, 2.3 percent; Libya, 2.0 percent; Brazil, 2.0 percent; Kazakhstan, 1.6 percent; Angola, 1.6 percent; and Qatar, 1.3 percent (Klare 2008:42). A research team at the University of Uppsala projects the peak of all petroleum liquids will occur by 2012 (see Gilbert and Perl 2010:119). Nonconventional means of producing oil include heavy, deepwater, and polar production. Heavy oil is derived primarily from tar or oil sands in northern Alberta, Canada. Deepwater oil is extracted at ocean depths of over 500 meters, primarily in the Gulf of Mexico and the South Atlantic. Polar oil is obtained from Arctic regions of the United States, Canada, and Russia. Flaring in the Nigerian Delta, according to a World Bank report, contributed more

Table 3.9. World's Top 10 Oil Producers, Actual (2005) and Projected (2030) Production

		Actual (2005)		*Projected (2030)*	
Rank	*Country*	*Million Barrels per Day*	*Percentage of World Total*	*Million Barrels per Day*	*Percentage of World Total*
1	Saudi Arabia	10.7	13.1	16.4	15.3
2	Russia	9.5	11.6	11.5	10.7
3	United States	8.0	9.8	9.1	8.5
4	Iran	4.2	5.1	5.0	4.7
5	China	3.8	4.6	3.3	3.1
6	Mexico	3.8	4.6	3.5	3.3
7	Norway	3.0	3.7	1.4	1.3
8	United Arab Emirates	2.8	3.4	4.9	4.6
9	Venezuela	2.8	3.4	1.7	1.6
10	Nigeria	2.8	3.4	5.2	4.9

Source: Adapted from Klare (2008:42).

greenhouse gas emissions than all other sub-Saharan African countries by 2002 (Zalik 2008:45).

Table 3.10 depicts the oil consumption of the top 10 oil-consuming countries in the world.

Table 3.10. Oil Consumption of the Top 10 Oil-Consuming Countries in the World, 2007

Rank	*Country*	*Barrels per Day (million)*
1	United States	20.68
2	China	7.578
3	Japan	5.007
4	Russia	2.858
5	India	2.722
6	Germany	2.456
7	Brazil	2.372
8	Canada	2.371
9	Saudi Arabia	2.311
10	South Korea	2.214

Source: Data from NationMaster.com.

Oil presently supplies about 40 percent of the world's energy and 96 percent of its transportation energy (Forest and Sousa 2006:1). The Institute for the Analysis of Global Security projects that world oil consumption will increase by about 60 percent between 2006 and 2020, and the US Energy Intelligence Administration projects an increase of over 50 percent between 2006 and 2025. The United States imports 60 percent of its oil, amounting to some 13 million barrels a day (Burman 2007:26). This figure is projected to rise 75 percent by 2030. In 2005, US oil imports came from the following places: 16 percent from Canada, 14 percent from Saudi Arabia, 12 percent from Mexico, 11 percent from Venezuela, 10 percent from Nigeria, 4 percent from Iraq, 3 percent from the United Kingdom, 3 percent from Russia, 3 percent from Angola, 3 percent from Algeria, and 14 percent from other countries (Burman 2007:28).

Of all the countries in the world, as none other than George W. Bush noted, "America is addicted to oil." In a more serious vein, Klare (2003:16) observes,

> Addressing the energy crisis was seen by Bush and his advisors as a critical matter for several reasons. To begin with, energy abundance is essential to the health and profitability of many of America's leading industries, including automobiles, airlines, construction, petrochemicals and agriculture, and so many shortages of energy can have severe and pervasive economic repercussions. Petroleum is especially critical to the US economy because it is the source of two-fifths of America's total energy supply—more than any other source—and because it provides most of the nation's transportation fuel. In addition to this petroleum is absolutely essential to US national security, in that it powers the vast array of tanks, planes, helicopters, and ships that constitute the backbone of the American war machine.

Oil serves as an important energy source in transportation, the production of asphalt, the heating of dwelling units and a wide array of other types of buildings, the fueling of mechanized farm equipment for industrial agriculture, and the production of pesticides. It is estimated that nearly half of global oil consumption is devoted to the products of the global auto industry. Oil is also used in the production of asphalt, plastics, and chemicals of various sorts.

Despite much discussion of the world approaching "peak oil," or even having already reached it, the oil industry continues to seek new sources of oil, regardless of the environmental consequences. This sad reality was poignantly illustrated by the Deep Horizon oil spill in the Gulf of Mexico in 2010. Despite continuing spillage in June 2010, the US government's Minerals Management Service granted BP new leases for deepwater drilling (Butler 2010a).

Farther north in North America, Imperial Oil, a subsidiary of ExxonMobil Canada, has been exploring the extraction of oil from the tar sands north of Edmonton, Alberta, which reportedly "releases at least three times the CO_2 emissions of regular oil production procedures and will likely become North America's single largest industrial contributor to climate change" (Jami 2010:16). Reportedly, about 1 billion cubic feet of natural gas are required to produce 1 million barrels of synthetic oil from tar sands (Klare 2008:41). Some experts contend that the tar sands area contains 1.7 trillion barrels of oil, which more or less matches the world's currently known reserves of conventional oil. China is increasing its access to oil. In May 2009, PetroChina, with its 75 projects in 29 countries, became the "world's largest traded company by capitalization" (Gilbert and Perl 2010a:304). Sinopec, another Chinese energy company, has constructed oil and natural gas pipelines from the Bay of Bengal to southwestern China. Shell, Chevron, and various small oil companies are exploring extraction of unmatured oil from oil shale, which exists in abundance in the Colorado River Basin of the western United States (Sperling and Gordon 2010:127). As Stuart Sim (2010b:179) so astutely observes,

> It is distinctly worrying that several potential new sources of oil have been identified recently such as under the Arctic Ocean, and that the prospect of being rescued from peak oil has meant that the environmental consequences of exploiting these has been sidelined in most public discussion so far. It is

also worrying that the more new oilfields that are brought on stream then the lower the price of oil is likely to be, which can only encourage reckless consumption.

Natural Gas

Natural gas was used as early as 1821 to heat homes in Fredonia, New York (White 2008:120). It began to take off as a major source of fuel in the 1880s, starting out with its use for lighting and industrial heat in Pittsburgh, Pennsylvania. As peak oil quickly approaches, if it has not already been surpassed, many parties view natural gas as a cleaner energy resource because it produces lower emissions than either oil or coal. It can be converted into many other products, such as liquid fuels, artificial fertilizers, and hydrogen for use in fuel cells. Table 3.11 depicts the world's top 10 holders of gas reserves. Iran, Qatar, and Russia contain 56 percent of the world's natural gas supply. The next five top holders of natural gas reserves are Kazakhstan, with 1.7 percent; Norway, with 1.6 percent; Turkmenistan, with 1.6 percent; Indonesia, with 1.5 percent; and Australia, with 1.4 percent. Fifteen countries contain 83.5 percent of world's natural

Table 3.11. World's Top 10 Holders of Natural Gas Reserves and Natural Gas Producers

		Reserves		*Production*	
Rank	*Country*	*Trillion Cubic Feet*	*Percentage of World Total*	*Billion Cubic Feet*	*Percentage of World Total*
1	Russia	1,682.1	26.3	21,607.1	21.3
2	Iran	933.0	15.5	3,706.5	3.7
3	Qatar	895.2	14.0	1,747.4	1.7
4	Saudi Arabia	249.7	3.9	2,601.6	2.1
5	United Arab	214.0	3.3	1,673.2	1.6
6	United States	209.2	3.3	18,500.7	18.5
7	Nigeria	189.9	2.9	995.5	1.0
8	Algeria	159.0	2.5	2,982.9	2.9
9	Venezuela	152.3	2.4	1,013.1	1.0
10	Iraq	111.9	1.7	843.7	0.8

Source: Adapted from Klare (2008:47).

gas reserves. Natural gas is used primarily to generate electricity and to heat homes and commercial buildings and for various industrial and agricultural purposes.

The United States is the world's leading consumer of natural gas, using 22.2 percent of this resource worldwide (Klare 2008:43). The European Union and Japan have been shifting from coal to natural gas in electricity generation so as to comply with the Kyoto Protocol (Klare 2008:44). Whereas oil is relatively easy to move by various means, including pipes, ships, trains, and trucks, natural gas is only easy to move by pipelines. Natural gas can be liquefied through a complex and expensive process and then transported in liquid natural gas (LNG) tankers, which pose enormous safety risks.

Steel, Aluminum, and Cement Production

The manufacture of products obviously too numerous to list contributes to greenhouse gases, particularly CO_2. Steel and aluminum are the most common metals utilized in the manufacture of numerous products, including motor vehicles, trains, airplanes, and ships, along with factories, office and residential buildings, appliances, and electronic equipment. Table 3.12 depicts world steel and aluminum production for selected years. In 2006 China was the leading steel producer, with 419 million tons, or over one-third of the world total (Liu 2008:56). The number two and three steel producers in 2006 were Japan, with 116 million tons, and the United States, with 99 million tons. Russia came in at number four and South Korea at number five in terms of steel production in 2006. Recycled iron and steel scrap has become an important raw material. In 2006 China accounted for 26 percent of the primary world aluminum production, and China, Russia, Canada, the United States, and Australia accounted for 59 percent of primary world aluminum production (Gardner 2008:58). While aluminum production accounts for about 3 percent of global electricity consumption, it accounts for some 10 percent of Australia's electricity consumption, which by and large is supplied by coal-fired power plants.

Table 3.12. World and Steel Production for Selected Years, 1950–2005

Year	*Steel Production (million tons)*	*Primary Aluminium Production (million tons)*	*Secondary or Recycled Aluminium Production (million tons)*
1950	190	1.5	0.4
1955	271	3.1	0.6
1960	347	4.5	0.9
1965	451	6.3	1.5
1970	595	9.7	2.2
1975	644	12.1	2.8
1980	716	15.4	3.9
1985	719	15.4	4.4
1990	771	19.3	5.8
1995	752	19.7	7.0
2000	847	24.4	8.2
2005	1,129	31.9	—

Sources: Data from Liu (2008:57) and Gardner (2008:62).

Cement and concrete are important components of many building materials and road construction and result in large amounts of CO_2 emissions. Chris Goodall (2010:229–30) observes,

> Nobody is quite sure, but most estimates suggest that the world cement industry is responsible for about 5 percent of global emissions. This surprisingly large figure arises because the world is now producing about 2.5 billion tonnes of cement a year, almost a third of a tonne per person on the planet. Much of this arises because of the pace of construction in China and other newly industrializing countries. But, to be blunt, Westerners bear some responsibility for this because we are buying the Chinese products made in the new buildings that are using so much cement and concrete.

Transportation

Metz (2010:153) reports that transportation accounted for 20 percent of total energy use in 2006, almost all of it in the form of oil products. Of this, motor vehicles accounted for more than 75 percent, with automobiles accounting for 45 percent, trucks 25 percent, and buses 8 percent. Airplanes, shipping, and rail transport accounted for 20 percent of the transportation energy consumption. According to Metz (2010:153), greenhouse gas emissions from the transport sector accounted for about 13 percent of the total global emissions in 2004. Paul Mees (2010:38), a transport expert at the Royal Melbourne Institute of Technology, presents an even grimmer picture of the impact of transportation on climate change:

> Transport is a critical element of the climate challenge, since it is the second-largest source of energy-related greenhouse gas emissions after electricity generation, accounting for around a quarter of the world total. Transport is also the fast-growing source of emissions, the rate of increase having overtaken that for electricity generation in the last decade. Three-quarters of transport-related emissions comes from road vehicles. [Organization for Economic Cooperation and Development (OECD)] countries currently account for two-thirds of global transport emissions, but developing nations are fast catching up.

Table 3.13 depicts changes in greenhouse gas emissions for selected countries from transportation and from all sources (except land use and forests) between 1990 and 2004. In 2005, in terms of passenger kilometers travelled per capita GDP (PKT/cap), in purchasing power parity, for various travel modes, North America came in at 36,263 PKT/cap kilometers, Western Europe at 22,302, Eastern Europe at 7,200, and developing countries at 3,660.

Motor Vehicles

Motor vehicles, with their internal combustion engines, alongside military aircraft and nuclear warheads, embody the social structural contradictions of the capitalist world system. In their monumental book *Monopoly Capital* (1966), Paul Baran and Paul Sweezy identified the steam engine, the railroad, and the automobile as having had "epoch-making" impacts on capitalist

Table 3.13. Changes in Greenhouse Gas Emissions for Selected Countries, 1900–2004, from Transportation and from All Sources (Except Land Use and Forests)

Country	*All Sources (%)*	*Transport (%)*
Australia	25.1	23.4
Bulgaria	–49.0	–32.1
Denmark	–1.1	26.8
France	–0.8	20.8
Germany	–17.2	5.1
Ireland	23.1	143.8
Italy	12.1	27.6
Japan	6.5	19.8
New Zealand	21.3	61.6
Spain	49.0	77.3
Sweden	–3.5	9.0
United Kingdom	–14.3	12.5
United States	15.8	28.1

Source: Adapted from Gilbert and Perl (2010:174).

development in their respective eras. They argued that the petroleum industry in large part has been a creation of the automobile. As Stilwell (1992:35) observes, the "private automobile serves the profit interests of the automobile, oil, tyre and ancillary industries, but it is far from being the best form of urban transport in terms of cost, health and ecological sustainability."

The world now has an estimated 700 to 800 million cars, and this number continues to grow rapidly as certain developing societies, such as China and India, imitate developed societies in their adoption of a culture of automobility (Montgomery 2010:9). According to Kovel (2007:71),

> Looming overcapacity hangs over automobile industries, as it does for capitalist production in general, with the ability to make some 80 million cars a year, and but 55 million or so able to be sold. Those unrealized 25 million vehicles are a giant splinter in the soul of capitalism, and the goad to endless promotion of automobilious values.

The United States is the largest producer of motor vehicles in the world, exemplified by the fact that in 2004 alone it manufactured 11.96 million (Rutledge 2006: xi). It is also the mostly highly motorized nation in the world, with 834 registered vehicles per 1,000 people, over 50 percent higher than Western Europe (Rutledge 2006:13).

China is the most dramatic instance in the growth of motor vehicles in the developing world. Diamond (Diamond 2005:36) reports,

> The number of motor vehicles (mostly trucks and buses) increased 15-fold between 1980 and 2001, cars 130-fold. In 1994, after the number of motor vehicles had increased 9 times, China decided to make car production one of its four so-called pillar industries, with the goal of increasing production (now especially of cars) by another factor of 4 by year 2010. This would make China the world's third largest vehicle manufacturing country, after the U.S. and Japan.

China in 2003 had around 24 million motor vehicles as opposed to the United States, which had about 250 million. According to Sachs (2008:76), "China's annual production is now soaring, up to around 7 million per year as of 2006 compared with just 2 million in the year 2000." In 2002, 1.2 million passenger vehicles were sold in China, and the following year this figure had grown by 40 percent (Rutledge 2006:135). Vehicle production in China declined during the last eight months of 2008 as a result of the Global Financial Crisis but then "rose steeply enough to resume or even exceed the pre-2008 trend" (Gilbert and Perl 2010a:303). The Chinese motor vehicle industry employs some 1.7 million workers (Klare 2008:70). Given that only 5 percent of Chinese households owned a car in 2009, the potential for an increase in the number of cars in China is huge (Montgomery 2010:8). Indeed, various studies project that China will have about 200 million cars by 2020 and nearly 400 million by 2030 (Montgomery 2010:37).

China has been feverishly building freeways both in the countryside and cities with over 21,000 miles (32,000 kilometers) traversing the country in 2004 and a doubling of this network by 2020 (Sperling and Gordon 2010:209–10). While freeways in the countryside still tend to be relatively lightly travelled, with a heavy concentration of trucks and buses, those in the cities, particularly large cities such as Beijing and Shanghai, have quickly become congested with all sorts of motor vehicles. In contrast, the Chinese government has not done much to develop commuter train and subway systems to accommodate its rapidly growing urban areas.

In India, although cars account for less than 10 percent of all passenger kilometers and rail and bus for about two-thirds, cars are increasing by about 30 percent per year (Metz 2010:155). John Farndon reports that "carmakers are expecting that India will soon be the seventh largest market for cars in the world, with 2.5 to 3 million new cars being bought each year." Whereas in 1951 India had an estimated 300,000 cars, this number had increased to about 85 million by 2005 (Shiva 2008:52). India's prime minister announced in 2007 a 10-year plan to transform his country into a "global hub of vehicle manufacture, design, and component production with a turnover of $145 billion by 2016," which would include special automotive zones in Chennai, Mumbai, and Kolkata (Shiva 2008:37). General Motors (GM), Honda, Volkswagen, and other car companies plan to build new factories in India, and Fiat, Nissan, and Renault are forming partnerships with Indian car manufacturers (Shiva 2008:50). Tata launched the minicar Nano in 2008 as an alleged "people's car" and has projected producing 1 million of them per annum by 2011. According to Shiva (2008:55), "Although the Nano gets good gas mileage, it is a fossil-fuel driven car," which "at 1 million new cars a year [will] contribute heavily to greenhouse gas emissions." Tata plans to market the Nano in Southeast Asia, Latin America, and Africa (Montgomery 2010:37).

Even in Singapore, which reportedly has an excellent public transport system, many affluent people are being seduced by the culture of automobility. According to Mees (2010:45),

> Car use in Singapore is growing rapidly. Mileage per vehicle is very high: the average car in Singapore travelled 21,100 km in 2006, double the figure for 1980 and virtually identical to the 21,317 km reported for Los Angeles. . . . The number of Singapore residents rose by 6 percent in the five years to 2007, but

> the number of cars jumped 19 per cent, traffic entering Singapore's CBD grew by 14 per cent and CBD traffic speeds fell by up to 30 per cent. Public transport's share of peak hour travel in 2004 was a very high 63 per cent, but still lower than the 67 per cent recorded in 1997. Car ownership is only 126 vehicles for every 1000 residents, but this is double the 1980 rate of 64, and continues to rise as income rises.

The International Monetary Fund projects the existence of some 3 billion cars in the world by 2050 (Richter 2010:118). Sperling and Gordon (2010:4) project there will be over 2 billion motor vehicles, at least half of them cars, by 2020. Other motor vehicles include trucks, buses, motorcycles, scooters, and electric bikes. Table 3.14 depicts the dramatic increase in motor vehicle production around the world following World War II. Automobile production has been concentrated in Europe, Japan, and North America, with production and utilization on the rise in developing countries. Thailand has evolved into the "car capital" of Southeast Asia with many major foreign automobile companies having manufacturing facilities there. The Chinese automobile industry consists of state companies as well as a number of joint operations between these companies and foreign companies, including Volkswagen, Toyota, Nissan, Honda,

Table 3.14. World Vehicle Production, 1950–2005, in Millions

Year	*Passenger Cars*	*Light Trucks*
1950	8.0	—
1955	11.0	—
1960	12.8	—
1965	19.0	—
1970	22.5	—
1975	25.0	—
1980	28.6	—
1985	32.4	—
1990	36.3	—
1995	36.1	12.9
2000	41.3	15.9
2005	45.9	18.5

Source: Adapted from Renner (2008:67).

Table 3.15. Motor Vehicles Manufactured in 2003 and 2004 in World Regions and Selected Countries

Country or Region	*2003*	*2004*	*Percentage Change*
Europe	20,000,286	20,829,774	4
North and South America	18,280,312	18,826,944	3
United States	12,114,971	11,089,387	–1
Brazil	1,827,791	2,210,062	21
Asia-Oceania	21,986,694	24,086,520	10
Australia	413,261	411,406	0
China	4,443,686	5,070,527	14
India	1,161,523	1,511,157	30
Japan	10,286,218	10,511,518	2
Africa	395,933	422,017	7

Source: Adapted from data from the International Organization of Motor Vehicle Manufacturers.

Hyundai, and General Motors (Dicken 2003:396–97). Table 3.15 provides statistics on the number of motor vehicles manufactured in various world regions and selected countries in 2003 and 2004.

Motor vehicles have had major impacts upon patterns of consumption, settlement (e.g., urban sprawl), traffic congestion, mass transportation, social relations, public policy, the environment, and health. Motor vehicles are a major contributor not only to air and noise pollution but also to greenhouse gas emissions. Numerous cities in the developed world and increasingly the developing world have evolved into what Peter Newman and Jeffrey Kenworthy (1999:31–33) term *automobile cities*. According to Peter Dicken (2003:359),

> In the Americas, both Canada and Mexico are tightly enmeshed with the US automobile industry . . . while Brazil remains the major automobile production centre in Latin America. The most striking new development of recent years has been the sudden emergence of South Korea as an important producer. As recently as the early 1980s, Korea was producing only 20,000 automobiles; in 2000 Korean output was 2.4 *million* (6.4 percent of the world total).

In a similar vein, Michael Cahill (2010:35) observes that "we see newly emerging consumer societies such as China and India promoting the car at the expense of the bicycle." Ironically, many cyclists probably find themselves so endangered by the growing onslaught of cars and even buses, jitneys, motor scooters, and motorcycles in developing countries that they feel forced to capitulate to motorized vehicular transport.

Car dependency is particularly pronounced in North American and Australian cities. Table 3.16 depicts the various modes of transport that people use to travel to work in various cities in these two regions. While North American and Australian cities vary widely in terms of the quality of their public transport systems, except in New York, the overwhelming majority of people in them opt to travel to work by car.

The Environmental Defense Fund (2007) released the following sobering statistics on motor vehicles and their contribution to greenhouse gas emissions in the United States alone:

- There are 232 million registered vehicles.
- The average US car consumes 600 gallons of gasoline per year.

Table 3.16. Methods of Travel to Work in Selected North American and Australian Cities

City	*Car (%)*	*Public Transport (%)*	*Walking (%)*	*Cycling (%)*	*Other (%)*
Los Angeles	91.1	4.7	2.7	0.6	1.1
Toronto	71.1	22.2	4.8	1.0	0.9
San Francisco	84.2	9.7	3.4	1.1	1.4
New York	20.5	67.6	24.8	0.3	1.6
Vancouver	74.4	16.5	6.3	1.7	1.1
Melbourne	79.3	13.9	3.6	1.3	1.9
Phoenix	93.4	1.9	2.1	0.9	1.4
Detroit	95.3	1.7	1.8	0.2	0.5
Canberra	82.0	7.9	4.9	2.5	2.7
Boston	85.1	9.0	4.2	0.4	0.9

Source: Adapted from Mees (2010:60–61).

- The average US car emits 12,000 pounds of CO_2 each year.
- US cars and light trucks travelled 2.7 trillion miles in 2004.
- Thirty percent of the world's automobiles are situated in the United States.
- The United States accounts for 45 percent of the world's automotive CO_2 emissions.

Cities vary greatly in terms of CO_2 emissions and other motor vehicle pollutants. Whereas the transportation-produced CO_2 in the New York metropolitan area totaled 3,378 kilograms per capita in 1990, it was 5,193 kilograms per capita in the same year in the Houston area (Newman and Kenworthy 1999:120). In contrast, Toronto emits 46 percent less CO_2 per capita than the average US city, largely due to an extensive public transportation system.

Despite improvements in US automobile fuel economy and emissions controls standards, a doubling of miles driven during the 1980s and 1990s by and large negated the impact of these innovations (Gonzalez 2008:164–65). Furthermore, while the catalytic converter "effectively breaks down the various nitrous oxides [NO_x] that contribute to smog and local air pollution . . . it creates nitrous oxide [N_2O], benign in smog creation but 300 times more potent than carbon dioxide as a greenhouse gas" (Porter 1999:81).

In *Autophobia: Love and Hate in the Automotive Age* (2008), Brian Ladd explores the contradictory nature of cars in the modern era. They have served to connect once isolated rural people and have been viewed as a form of "freedom"; cars have not only shaped much of modern life but constitute the "greater contributor to what we are told is our 'unstable lifestyle'" (Ladd 2008:6). Ladd (2008:8) argues that with automobile technology "powerful interests have made their influence felt in lasting ways, notably in the construction of roads." Furthermore, the automotive industry stimulated growth in other industries, including steel, rubber, glass, plastics, upholsteries, electronics, and tourism. The automobile also has contributed to urban and suburban sprawl. Automobile companies in their advertisements often depict cars in natural, often remote settings, such a mountain tops, canyon lands, and secluded beaches. As Low et al. (2005:234) observe, "Nature, like sex, sells cars! And yet nothing despoils nature like cars!"

The products of the global automobile industry account for nearly half of global oil consumption (Dauvergne 2005:43). The car consumes up to 63 percent of the oil used in the United States and about 35 percent of the oil consumed in Japan, a country with a vastly superior public transportation system. Oil is also a major resource utilized in road construction (Patterson 2007:38). Automobiles require an inordinate amount of fuel, a demand that is spurred on by the growing demand in developed societies for bigger and bigger vehicles. For example, the sports utility vehicle (SUV) market share in the United States increased from 2 percent in 1975 to 24 percent in 2003 (Leggett 2005:22). Despite the fact that political scientist Robyn Eckersley (2006:279) has dubbed Sweden a working prototype of the "green state," it reportedly overall has the highest-pollution-emitting cars, particularly Volvos and Saabs, in Western Europe (Ekan 2007).

Around the world, nearly 4,000 square kilometers of land, much of it farmland, are transformed annually for motor vehicle use in the form of roads, highways, and parking lots (Dilworth 2010:362). According to Shiva (2008:51),

> The car has seriously divided India. People can no longer walk on the streets. Neighbors have turned into enemies over car parking. It has cut up rural India through land grabs for factories and highways.

Small motorized transport such jitneys, scooters, and motorcycles have become a very popular mode of transport among middle-income individuals, particularly in developing countries, and obviously contribute to greenhouse gas emissions.

The Global Financial Crisis resulted in a downturn in automobile sales with a 30 percent reduction in late 2008 over the previous year (Barbier 2010:79). General Motors and Chrysler declared bankruptcy in the United States with GM receiving a US government bailout. Despite this, worldwide automobile utilization can be expected to rise in coming decades.

Airplanes

While the German carrier Lufthansa pioneered civilian air travel in the 1920s, flying was by and large an elite mode of transport until the 1950s. According to Wolfgang Behringer (2010:178),

> Only when it became normal practice in the business world did it come down in price and become accessible to much wider layers. The arrival of mass tourism in the 1970s brought a rapid increase in the numbers of passengers and destinations as well as the capacity of individual aircraft. Energy consumption per passenger rose disproportionately, however, as an aircraft holding three hundred needs as much fuel as tens of thousands of Volkswagen Beetles.

Table 3.17 depicts the profound increase in world air travel during the period from 1950 to 2005. While motor vehicles are a major contributor to greenhouse gas emissions, various sources indicate that air travel alone may be contributing from 3 to 8 percent of greenhouse gas emissions (Spence 2005:148; www

Table 3.17. World Air Travel by Distance and Passenger Volume, 1950–2005

Year	*Passengers (million)*	*Distance (billion passenger kilometers)*
1950	31	28
1955	68	61
1960	106	109
1965	177	198
1970	383	460
1975	534	697
1980	748	1,089
1985	899	1,367
1990	1,165	1,894
1995	1,304	2,248
2000	1,674	3,038
2005	2,022	3,720

Source: Adapted from Chafe (2008:71).

.chooseclimate.org/flying). Unfortunately, the Kyoto Protocol exempts emissions from aviation and marine shipping.

Airplane travel presently represents 12 percent of CO_2 emissions from transport (Gautier 2008:118). Airplanes also emit nitrous oxide and other contrail or exhaust fumes, meaning that a "factor between two and three is normally applied to the CO_2 emissions from aviation to account for the additional warming impact" (Tickell 2008:41). On the positive side, Schaefer et al. (2009:157–58) project that technological innovations will result in more energy-efficient airplanes. For the Boeing 777, based on their calculations, they anticipate that

> new aircraft energy use per seat kilometer is expected to decline by roughly 25–45 percent by the mid 2020s. When the projected improvements in air traffic management are included, aircraft energy intensity declines by roughly 30–50 percent by the mid 2020s. This is equivalent to a 1.2–2.3 percent per year reduction in energy intensity, compared with an average rate of 3.2 percent per year over the past thirty years.

Various projections indicate a continuing rise in air travel. The 2007 Intergovernmental Panel on Climate Change report projects that air travel emissions will account for up to 15 percent of greenhouse gas emissions by 2050. Although the UK government has set a target of reducing CO_2 emissions 80 percent by 2050, its 2003 white paper on aviation projects an increase from 200 million passengers moving through UK airports at the present time to as many as 400 million by 2020 and 500 million by 2030 (Cahill 2010:5). To facilitate this projected increase, it has approved airports, taking measures to expand facilities to be able to handle the increased number of flights.

As in many other areas of a stratified world system, the affluent contribute much more overall to greenhouse gas emissions from flights than working-class people and particularly the poor around the globe. In the case of one developed society, the United Kingdom, the Oxford Transport Studies Unit confirmed this grim reality in its research:

> Although over half of the UK population now travels by air at least once a year—though almost half do not—a very small percentage of people travel many times as often. . . . Cheap air travel may seem to be a great leveller, making long-distance travel available to all; but its most important impact has probably been to allow the richest few per cent of the population almost unlimited freedom to pollute as much as they want, barely thinking about the financial impact. (Cited in Goodall 2010:183–84)

Given the present predilection for ongoing corporate growth and globalization, what type of future can we foresee for airplane travel and airports as their hubs? John Kasarda, at the University of North Carolina, Chapel Hill, has coined the term *aerotropolis* for what he foresees as a world of new cities built around airports that interconnect with numerous other *aerotropoli* around the world, just as seaports and railways served as urban hubs in the past (Kasarda and Lindsay 2011). The Persian Gulf region of the Arabian Peninsula has evolved into the hub of several aerotropoli, such as Dubai and Abu Dhabi in the United Arab Emirates, which, according to Kasarda and Lindsay (2011:309), "expects to receive 280 million passengers a year by 2015."

Marine Transportation

Marine transport has sometimes been posited as a reasonably sustainable sector for moving freight compared with airplanes, trucks, and railways. In terms of CO_2 emissions, freight transport by road creates 98.301 grams/ton kilometer; rail, 28.338 grams/ton kilometer; and short-sea shipping, 15.450 grams/ton kilometer (Black 2007:209). Unfortunately, in terms of the emissions of sulfur dioxide, short-sea shipping emits 0.290 grams/ton kilometer, as compared with 0.031 grams/ton kilometer for road transport and 0.036 grams/ton for rail transport. Catherine Gautier (2008:119) reports that "in 2000 EU-flagged ships emitted almost 200 [metric tons (Mt)] CO_2, which is significantly more than from EU aviation sources."

The shipping of oil in tankers has also sadly resulted in numerous oil spills since 1967 (Black 2010:21). Like airplanes, ocean shipping is subject to the availability of fuel in the era of peak oil and the economic condition of the capitalist world system (e.g., the Global Financial Crisis). According to Bill McKibben (2010:89),

> By May [2008], the cost of sending a ship container from Shanghai to the United States was eight thousand dollars, up from three thousand dollars at the beginning of the decade. Cargo volumes began to fall.

Cleo Paskal (2010:80) describes sea shipping as the "circulatory system of the global economy" in which about "90 percent of the world trade products are carried at some point." Sea shipping increased between the early 1960s and 2006, going from less than 6 trillion ton-miles to 33 trillion ton-miles. The cost of shipping has decreased by about 80 percent in 25 years due to "containerization, bigger ships, and computer-assisted resource allocation" (Paskal 2010:81).

Dwelling Units and Buildings

According to Amanda Little (2009:319), "Buildings alone account for nearly 40 percent of all energy use and contribute nearly 40 percent of the world's annual greenhouse gas emissions." Table 3.18 depicts primary energy use in residential and commercial buildings in the United States.

Dwelling units in developed societies, particularly in North America, Australia, and New Zealand, have become larger and larger. In the case of Australia,

> between 1985 and 2000 the average floor area of new homes increased by 31 per cent, from 170 square metres to 221 square metres, and the size of apartments increased by 25 per cent, to an average of 139 square metres. In the mid-1950s, the average size of new houses was about 115 square metres, that is, half the size of houses today. (Hamilton and Denniss 2005:20)

Table 3.18. Primary Energy Use in Buildings in the United States

Residential	*Percentage*	*Commercial*	*Percentage*
Space heating	32	Lighting	27
Space cooling	13	Space heating	15
Water heating	13	Space cooling	14
Lighting	12	Water heating	7
Refrigeration	8	Electronics	7
Electronics	8	Ventilation	6
Cooking	5	Refrigeration	4
Wet cleaning	5	Computers	3
Computers	1	Cooking	2
Other	3	Other	15
Total energy	**21.8 quads***		**17.9 quads**

*A quad is a million billion British thermal units.

Source: Adapted from Richter (2010:112).

These larger dwelling units generally require more and more energy to heat or cool and demand that their occupants purchase more and more consumer items, such as wide-screen television sets, computers, huge BBQ grills, and so on. For example, since the 1990s, many Danes reportedly have built their bathrooms to include "more toilets, double sinks, vanities, spa-like tubs, independent showers with surround spray, and more" (Montgomery 2010:28).

There has been a growing tendency to overheat and overcool dwelling units and to rely upon central warming and cooling systems, depending upon the season. In the United Kingdom, reportedly "average winter house temperatures rose sharply from about 12 degrees (53.6°F) in 1970 and hit 18 degrees (64.4°F) across the whole house in about 2002" (Goodall 2010:42). In reality, many people now expect indoor temperatures during the winter to be a few degrees warmer, thus enabling them to move around or sit in short-sleeved shirts rather than bundling up, or "rugging up," with additional layers to keep warm. The average indoor temperature in Swedish homes during winter is currently around 21°C, and in German homes its a toasty 22°C

(Goodall 2010:69). Conversely, during the summer months, there often is tendency to air-condition to a temperature of around 68°F (20°C), much cooler than necessary for the vast majority of people.

A Seemingly Endless Array of Consumer Items

To survive, capitalism must generate an artificial need—namely, the need to consume endlessly a wide array of commodities, even potentially dangerous and lethal ones. As Costas Panayotakis (2006:265) observes, "As capitalist consumer culture continues to liquidate non-commercialized local cultures, the 'migration of meanings and values from relationships with people to relationships with market goods and spectacles' channels people's consumption preferences and conceptions of the 'good life' in a consumerist direction." Perhaps I can best illustrate this pattern by relating a personal anecdote that I experienced while doing a postdoctoral fellowship at Michigan State University.

While we were residing in Lansing in 1979 to 1981, my family's next-door neighbor was a relatively well-paid Oldsmobile assembly-line worker who lived with his schoolteacher wife and no children in a modest two-story house, more or less similar to our house. On three or four occasions, I asked him to tell me about his work at the Oldsmobile plant, but on each occasion he deflected our conversation to a discussion of the consumer items that he and his wife owned and apparently enjoyed immensely. Their home was filled with the latest electronic items of that era, and outside the house sat a large recreational vehicle (RV), probably a Winnebago. On one occasion he gave me a tour of his RV and told me about the RV club to which he and his wife belonged and the excursions that they made with fellow club members to various places on weekends and holidays. He proudly put on his RV club jacket and told me of how, upon arriving at an RV camping park, the club members would form a circle very much in the vein of pioneer covered-wagon days.

One of the families would be designated to purchase groceries at the nearest town for their festivities. I surmised that at least one reason why my neighbor did not wish to discuss his work was that he found it so routine, boring, and particularly alienating. Rather than in his 40 or more hours of work on the assembly line, he had found compensation, meaning, and community in a cornucopia of consumer goods that he shared with other consumers, particularly in his RV club.

Consumer capitalism began to take off in a profound way beginning in the 1950s when households in developed societies were flooded with energy-intensive appliances and devices, including electric cookers, washing and drying machines, refrigerators, toasters, electric irons, microwave ovens, electric toothbrushes, electric razors, television sets, record players, cassette players, video recorders, computers, printers, electric tools, power lawnmowers, hedge cutters, leaf blowers, elaborate lighting systems, and so forth. Capitalism, with its predilection for built-in obsolescence, encourages people to update older models with new ones, such as the latest plasma televisions, CD and DVD players, mobile phones, and many other items.

Pat Murphy (2008:232) argues that many electrical household appliances, such as garbage disposals, clothes dryers, toasters, and food processers, as well as lawn-care machines, electric hedge trimmers, lawn mowers, and leaf blowers, could be replaced with manual devices or altered practices, such as hanging clothes outside and using push mowers. A 2002 Rocky Mountain Institute report indicates that in the average US home, electronic appliances, including answering machines, VCRs, and stereo sets, draw electricity even when shut off, wasting 587 kilowatt hours or about 840 pounds of carbon a year (Charman 2008:35). According to T. Brennan (2003:6), "Computers have drastically increased the use of coal and in this they are typical communications products, relying on electricity which in turn relies on fossil fuels."

Refrigeration and air-conditioning (including in motor vehicles) account for some 80 percent of the utilization of the F-gas family of chemicals—namely, the chlorofluorocarbons

(CFCs), hydrochlorofluorocarbons (HCFCs), perflurocarbons (PFCs), and sulfur hexafluoride (SF_6) (Mate, Davies, and Kanter 2009:52). These chemicals are also used as "solvents, blowing agents in foams, as aerosols or propellants, and in fire extinguishers" (Mate, Davies, and Kanter 2009:52). The chemical industry developed HCFCs as a substitute for CFCs that damage the ozone layer, but the former are a class of potent greenhouse gases. Reportedly, environmentally friendly alternatives to F-gases are available for almost all domestic and commercial applications.

Computers require a tremendous amount of material and energy resources both in their production and operation. Reportedly, "one desktop computer and monitor, averaging fifty-three pounds, requires at least ten times its weight in fossil fuels and chemicals" (Rosa and Dietz 2010:29). Furthermore, computers generally have a short life span, in large part due to upgrading to newer models, resulting in the disposal of outdated models that contain toxic materials that often become e-waste. The computer industry is notorious for the built-in obsolescence of its products and its pattern of encouraging users to upgrade their computers frequently.

Last but not least, the manufacture of cell or mobile phones and the operation of a mobile phone network result in a significant amount of greenhouse gas emissions. For the United Kingdom, Goodall (2010:225) provides the following tabulation of greenhouse gas emissions emanating from a single cell or mobile phone per annum: 24 kilograms for its manufacture, 11 kilograms for its operation within the network, 2.4 kilograms for leaving it in a charger, 0.3 kilograms for the actual charging, and 0.04 kilograms for making calls. Of the some 30 million cell or mobile phones sold annually in the United Kingdom, most of them are manufactured overseas, meaning that their carbon cost does not appear in the country's national emissions accounts.

While developed societies constitute the leading cultures of consumption, various developing countries, such as China and India, are quickly joining the pack. Many of the refrigerators, air conditioners, washing machines, televisions, and computers manufactured in China are exported; yet, many of them are sold to the members of the new Chinese middle class as well as elites. Klare (2008:69) reports that most of the apartments and shopping malls in Shanghai are "cooled in summer by air-conditioning; most house computers and other advanced electronic devices as well as a wide variety of modern appliances [are] all powered by a vast electrical grid."

The capitalist culture of consumption also encourages people to purchase many other products. Worldwide each year, $18 billion is spent on makeup, $15 billion on perfume, $14 billion on ocean cruises, and $15 billion in the United States on Christmas presents for pets alone (Taylor 2008:70). Many consumer items, particularly electronic products, which contain toxic parts, are discarded in landfills that produce landfill gas consisting of about 50 percent methane (Sheehan and Spiegelman 2010:374). Even pets, which obviously may serve socially and emotionally useful roles (especially for isolated elderly individuals), are part and parcel of the culture of consumption. Eminent sociologist Zygmunt Bauman (2011:24) reports that "Europe and the United States spend 17 billion dollars each year on animal food, while according to some experts, just 19 billion dollars is needed to save the world's population from hunger."

Tourism and Travel

Since the advent of cheap air flights, tourism has skyrocketed, becoming an integral component of the culture of consumption. The World Travel and Tourism Council estimated tourism to be an $8 trillion industry, accounting for almost 10 percent of global GDP in 2008, and projected that this figure would rise to $15 trillion by 2018 (Sim 2010a:132). Many countries and regions, such as the Caribbean, Europe, Latin America, Southeast Asia, Australia, New Zealand, and the islands of the South Pacific, rely very heavily upon tourism as a source of income. Tourism makes up 17.2 percent of Greece's GDP (Sim 2010a:136). While some tourists, such as backpackers, seek to travel simply and by train, coach, or ferry, many rent cars, or fly hither and thither on planes, or take luxury cruise ships and stay at luxury resort hotels that require tremendous energy and resources to operate. The World Tourism Organization and United Nations Environment Programme (2008) conducted an analysis of the CO_2 emissions generated by various components of tourism in 2005. The World Tourism Organization (2008:33) concludes that internationally emissions from transport, accommodation, and tourism represented "between 3.9% and 6.0% of global emissions in 2005, with the best estimate of 4.9%." The report indicates that the "total CO_2 emissions from tourism transport are estimated to be in the order of 980 Mt CO_2[;] 52% of these is estimated to be caused by air travel (515 Mt CO_2), 43% by car (420 Mt CO_2), and 5% by other forms of transport—coach, rail and water borne—(45 Mt CO_2)," coming to a total of 1,302 Mt CO_2 out of a total of 26,4000 Mt CO_2 (World Tourism Organization 2008:127). Total CO_2 emissions associated with tourist accommodations came to an estimated 274 metric tons, and other tourist activities made up 48 metric tons of CO_2 in 2005. It is difficult to determine the CO_2 emissions that result from other tourist activities, such as visiting museums, theme parks, sports events, musical and theater events, shopping, and visits to friends. Tourism and even ecotourism become another component of the culture of consumption in that people may collect sights and experiences, often permanently captured in the form of photographs and self-produced movies, as yet another consumer item. A former colleague of mine in the United States had a competition with a friend to see who had visited the most countries, even if only for a few days.

In the case of ecotourism, people "see the environment in simplified (a-social, a-historical and a-ecological) terms, which obscure the socio-ecological implications of the global infrastructure and economic relationships that make ecotourism possible in the first place" (Brockington, Duffy, and Igoe 2008:145). J. G. Carrier and D. V. I. Macleod (2005) draw attention to the reliance of ecotourism on airplane travel, which, as I have already noted, is growing as a contributor to greenhouse gas emissions. As Dan Brockington, Rosaleen Duffy, and Jim Igoe (2008:145) observe, "the ecological footprints of ecotourists who fly to different parts of the world on a regular basis, therefore, is usually several orders of magnitude more significant than the footprints of the local people who live in the places that ecotourists visit." Of course, other forms of travel related to business, including conferences of many sorts, and visits to relatives and friends also contribute to greenhouse gas emissions.

Industrial Agriculture and Logging

The Green Revolution and industrial agriculture have been reliant upon fossil fuels, extending from the production of fertilizers, to the operation of farm machinery, to the transportation and storage of agricultural products. An estimated three calories of fossil fuel energy are required to produce one calorie of food energy (Bello 2009:36). Reportedly 400 gallons of oil equivalents were needed in 1994 to feed each American for the year (Ruppert 2009:86). Michael Ruppert (2009:89) provides the following sobering account of the energy demands and environmental impacts of industrial agriculture:

> Technologically-enhanced agriculture has augmented soil erosion, polluted and overdrawn groundwater and surface water, and even (largely due to increased pesticide use) caused serious public health and environmental problems. Soil erosion, overtaxed cropland and water resource overdraft in turn lead to even greater use of fossil fuels and hydrocarbon products. More hydrocarbon-based fertilizers must be applied, along with more pesticides; irrigation water requires more energy to pump; and fossil fuels are used to process polluted water.

Roundwood production, which includes wood that is cut down from forests and other areas or simply retrieved from the forest floor, continues to increase, as table 3.19 indicates.

As indicated in table 3.20, deforestation, intensive tillage, and overgrazing release CO_2 from living or recently living plants and soil organic matter. Growing plants can sequester huge

Table 3.19. World Roundwood Production

Year	*Production (million cubic meters)*
1965	2,475
1970	2,644
1975	2,705
1980	2,978
1985	3,162
1990	3,382
1995	3,251
2000	3,358
2005	3,503

Source: Adapted from Gardner (2008:63).

amounts of CO_2 from the atmosphere and store it in vegetation and soils. Conversely, land changes contribute to the release of CO_2, nitrous oxide, and methane.

Table 3.21 depicts the total carbon budgets and CO_2 emissions from land-use changes and fossil fuel combustion for selected countries. As timber tends to be an export crop for

Table 3.20. Greenhouse Gas Emissions from Land Use (million tons CO_2e)

Land Use	*Annual Emissions*	*Greenhouse Gas Emitted*
Agriculture	6,500	
Soil fertilization (inorganic fertilizers and applied manure)	2,100	N_2O
Gases from food digestion in cattle (enteric fermentation in rumens)	1,800	CH_4
Biomass burning	700	CH_4, N_2O
Paddy (flooded) rice production (anaerobic decomposition)	600	CH_4
Livestock manure	900	CO_2, N_2O
Deforestation (including peat)	8,500	
Agriculture and livestock	5,900	
Total	15,000	

Source: Adapted from Scherr and Sthapit (2009:32).

Table 3.21. Total Forest Carbon Budget and CO_2 Emissions for Land-Use Change and Fossil Fuel Emissions, 2005

Country	*Total CO_2 Emissions (Mg CO_2 per capita)*	*Total CO_2 Emissions from Land Use (% of total)*
Australia	19.4	7
Canada	19.8	16
Japan	10.4	8
United States	17.7	0
Germany	9.5	0
United Arab Emirates	30.1	0
Mexico	4.9	17
Russia	12.4	16
Brazil	8.2	79
Venezuela	10.4	46
China	4.3	0
Peru	7.5	81
Indonesia	8.8	78
Nepal	2.8	96
Nigeria	1.6	49

Source: Adapted from Vogt et al. (2010:110).

most developing countries involved in logging, the resulting emissions are essentially export ones that do not show up in the emissions figures for the developed countries that are importing timber and other wood products.

Meat production requires 7 kilograms of grain for 1 kilogram of beef, 3.5 kilograms of grain for 1 kilogram of pork, 2 kilograms of grain for 1 kilogram of poultry, and 1.2 kilograms of grain for 1 kilogram of fish (Metz 2010:235). Around the world much soybean production is devoted to feeding livestock (Dauvergne and Lister 2011:128). In 2002, regional meat production per capita came to 271 pounds in North America, 154 pounds in South America, 62 pounds in Asia, 163 pounds in Europe, 103 pounds in Central America, and 57 pounds in North Africa (Murphy 2008:194). The UN Food and Agriculture Organization has indicated that meat production alone may be contributing to 18 percent of total global greenhouse gas emissions, largely due to the fact that livestock emit huge amounts of methane (Hertsgaard 2011:180). Peter Dauvergne and Jane Lister (2011:129) report,

> Global meat production has increased almost sixfold over the last 50 years. And it is on track to double again by the middle of this century. Already, there are more than 1 billion pigs, 1.3 billion cattle, 1.8 billion sheep and goats, and 15 billion chickens. . . . The US is the world's largest per capita consumer of meat, averaging over 125 kilograms, or 275 pounds a year. Beef is at the heart of this meat diet and the US is the world's largest beef consumer; however, growth rates are highest in developing countries with historically low per capita rates of production, particularly China.

A vegetarian diet reportedly requires about 80 percent less land than what is needed to feed a person on a meat-based diet (Metz 2010:46). According to Metz (2010:241), "Changing to a vegetarian diet can avoid N_2O emissions from grasslands, CH_4 emissions from livestock and manure, CO_2 emissions from fossil fuel use, and free land for other purposes." Even in those instances where animal production would continue, methane from animal digestion could be reduced through practices such as vaccinations and chemical inhibitors. Changes in tillage practice and vegetation cover and fertilizer management, soil and water management, and fertilizer additives could also reduce N_2O emissions from fertilizers (Rickards and Tucker 2009:93).

Many people in developing countries continue to rely on fuel wood for energy for heating, cooking, and other activities. According to Kristina Vogt et al. (2010:107),

> Today, half of the global population is still dependent on forests for traditional products they provided society several thousand years ago (woodfuel for cooking and heat). Many countries (e.g., Bangladesh, China, Haiti, Iceland, the Netherlands, Nigeria and the UK) have such low forest cover that they urgently need to plant trees to restore their forests to some semblance of what existed historically.

While various developing countries do not rely on wood and other biomass sources as much as they did in the past, biomass sources still constitute a major energy source that also emits carbon dioxide, as indicated in table 3.22.

Semiperipheral countries, such as Brazil and Indonesia, are the sites of the most intense logging activities or deforestation in the context of the capitalist world system. In the mid-1990s, Japan obtained up to 50 percent of its log imports and 98 percent of its plywood imports from Southeast Asia (Bartley and Bergesen 1997:369). According to Tim Bartley and Albert Bergesen (1997:369),

> Semiperipheral countries deforest more than others because of their position of potential upward mobility in the world-system, which leads them to place more weight on industrialization than on environmental protection. . . . Because of the *potential* for economic development, semiperipheral countries are more eager to reap the economic benefits of forest exploitation than are developed countries. Further, semiperipheral countries have a greater technological capability to deforest than peripheral countries.

The World Resources Institute reports that more than 60 percent of deforestation in the world occurs in Brazil, particularly

Table 3.22. Percentage of Total Energy Use Supplied by Fuel Wood in Selected Developing Countries

Country	*1980*	*1996*	*2003*
Bangladesh	81.3	43.3	51.5
Brazil	35.5	29.2	29.1
China	8.4	5.6	4.6
Ethiopia	89.6	93	96.5
Ghana	43.7	78.1	84.7
India	31.5	21.2	19.8
Nepal	94.2	90.9	93.2
Papua New Guinea	65.4	62.5	62.2
Zimbabwe	27.6	23.4	67.2

Source: Adapted from Potter et al. (2008:257).

in the state of Mato Grasso in the Amazon Basin, and Indonesia, particularly in Riau Province, which contains large peat forests.

Deforestation reportedly "increased Brazil's total carbon emissions fivefold in 2002, moving it from the ninth-largest emitter to the fourth-largest after the United States, China, and Russia" (Bodley 2008:37). According to Gore (2009:175), "While Brazil is destroying twice as much forestland each year as Indonesia, Indonesia is emitting twice as much CO_2 from deforestation as Brazil—primarily because the carbon-rich peatlands from which the Indonesian forests are being cleared dry up when the tree cover is gone and burn much longer when set ablaze, emitting far larger quantities of CO_2 into the atmosphere." Table 3.23 depicts the top 10 deforesting countries.

Over half of the world's paper is consumed by the wealthiest 20 percent of people; yet, much of the deforestation that produces this paper is concentrated in the semiperiphery and periphery (Synott 2004:224). Americans constitute the largest global per capita consumers of paper, much of which is used for advertisements in newspapers and magazines and "junk mail" (Dauvergne and Lister 2011:116). The Amazon Rainforest experienced widespread drought in 2005, which transformed the region from a sink to source of carbon (Allison et al. 2009:41).

Table 3.23. Top 10 Deforesting Countries

Country	*Rank*	*Net Loss of Forest Acres per Year, 2000*
Brazil	1	7,667,649
Indonesia	2	4,623,322
Sudan	3	1,455,445
Myanmar	4	1,151,506
Zambia	5	1,099,614
Tanzania	6	1,018,070
Nigeria	7	1,013,127
Democratic Republic of Congo	8	788,263
Zimbabwe	9	773,436
Venezuela	10	711,661

Source: United Nations Food and Agriculture Organization 2007.

Militarism and Wars

The contribution of military operations to global warming remains an underresearched topic. However, the Pentagon's activities reportedly resulted in about 46 million tons, or 3.5 percent, of US CO_2 emissions in 1988 (Renner 1997:121). More recent figures indicate that CO_2 emissions came to 60 million tons from US military operations in 2005 and 5 million tons from UK military operations in the same year (Parkinson 2007:4). According to Michael Renner (1997:121),

> Estimating a global figure for carbon emissions from the military is fraught with uncertainty. A back-of-the-envelope calculation for the late 1980s yields an estimate of about 150 million tons: almost three percent of the global total, or nearly equal to the annual carbon emissions of the United Kingdom. If the energy consumption of arms-producing industries were included, these numbers could well double.

Militaries, with their heavy reliance on airplanes (ranging from fighter jets, to planes carrying troops and cargo, to Air Force One), battleships, aircraft carriers, tanks, and other military equipment, rely heavily on oil. The 2007 *Energy Bulletin* reported that the Pentagon is the single largest consumer of oil in the world, with an official figure of 320,000 barrels of oil per day being used for vehicle transport and facility maintenance (Korbez 2007). The official figure does not include "energy for the manufacture of vehicles, energy for building and dismantling military facilities, energy for construction of roads, and energy consumed while rebuilding whatever the military blows up" (Fitz 2007:1). Klare (2007) maintains that the Pentagon consumed 134 million barrels of oil in 2005, as much as the entire country of Sweden. One-third or more of US military oil consumption reportedly occurs outside the United States (Smith 1990–1991:1). While the figures need updating, Gar Smith (1990–1991) reported that around 1990,

> the biggest gas-hogs in the Pentagon's arsenal [were] the Navy's non-nuclear aircraft carriers that burn 134 barrels per hour

> and battleships which consume 68 barrels per hour. At a speed of 25 knots, the USS *Independence* (a 1070-foot-long aircraft carrier with . . . a flight deck and a crew of 2300) consumes 150,000 gallons of fuel a day.

Linda McQuaig (2004:3) observes, "Even as the competition over dwindling reserves heats up and threatens to cause international conflict, we are faced with a still more devastating consequence of our addiction to oil—global warming. " While war is contributing to climate change vis-à-vis greenhouse gas emissions, the latter in turn may already be contributing to conflicts in drought-stricken regions of sub-Saharan Africa and threatens to pose larger-scale conflict as the twenty-first century unfolds. In anticipation of such a possible scenario, at the same time that the Bush administration was downplaying the seriousness of global warming, the Pentagon commissioned a report titled "An Abrupt Climate Change Scenario and Its Implications for United States National Security" (Schwartz and Randall 2003), which was not intended for public consumption. Figure 3.1 depicts a model of the impact of the capitalist oil–motor vehicle–military complex on war, climate change, and health.

In his discussion of *petro-imperialism,* Elmar Altavater (2006:51) argues that "oil security" constitutes one of the priorities of the United States and other powerful oil-consuming countries and blocs, such as the European Union. He delineates three strategies that these countries and blocs have utilized in their efforts to gain strategic control over oil: "Diplomacy, and the establishment of friendly inter-state relations, as in the Gulf region; or by means of subversion, as in some Latin American and African countries; or by using massive military power, such as in Iraq, and to a lesser extent also in Central America—and perhaps in the future against Iran and Venezuela." According to Vandana Shiva (2008:15–16), eco-imperialism, which entails the "control over the economies of the world through corporate globalization," includes the "oil wars being fought in the Middle East, Central Asia, and Africa and the new land and food wars triggered by the emergence of industrial biofuels."

Capitalism → Oil as a Vital Energy Resource → Military Machine & Wars for Oil

↓ ↓ ↓ ↓

Motor Vehicles and Other Transport Forms as Sources of Profit → Pollution & GHGEs

↓ ↓

Global Warming & Related Climatic Changes

↓ ↓ ↓

Conflicts Resulting from Global Warming → Injuries, Disease, & Death

Figure 3.1.

The need for more and more oil drives a global war machine that both pushes and is the force that allows countries to go after this scarce resource along with other scarce resources. Eventually, resource wars erupt in various regions, thereby creating scenarios in which the carbon-based military machine pollutes the environment, causes global warming, creates diseases, and kills and maims people. Environmental degradation, which includes global warming, in turn creates greater competition for resources and raises the specter of even more violent conflict, including war.

It is difficult to determine the amount of greenhouse gas emissions released during actual wars themselves. During the 1991–1992 Persian Gulf War, Iraq allowed over 600 oil wells to burn for a period, emitting an undetermined amount of CO_2 into the atmosphere. According to one source, "An estimated three to six million barrels of oil per day were burnt (compared with an average daily consumption of oil in all Western Europe of 12 million barrels)" (Castells 2004:214). Helen Caldicott (2009:14) reports that in the aftermath of the First Gulf War,

> 732 oil wells were burning in Kuwait, set alight or damaged by departing Iraqi soldiers and US weapons that had missed their targets. Red Adair and other oil well fire specialists were brought in to quell the fires, and estimates of the time for completion of the task varied from two to ten years. By mid-September 1991, some 429 of the smaller wells had been recapped. . . . Some scientists say that the carbon dioxide emissions produced in the four months after the war ended were equivalent to 1 percent the normal global production.

Barry Sanders (2009:22) asserts that the "military—that voracious vampire—produces enough greenhouse gases, by itself, to place the entire globe, with all its inhabitants large and small, in the most imminent danger of extinction." The Pentagon is the largest consumer of oil in the United States. In the War on Iraq, the US military used 1.5 million gallons of fuel every day to power its tanks, fighter jets, Black Hawks, Humvees, hospitals, and base camps (Little 2009:47). According to Amanda Little (2009:47–48), "In recent years [the Pentagon] has used between 130 million and 145 million barrels of oil annually—comprising 2 percent of America's total petroleum." Ironically, as she goes on to observe, "for the better part of a century, oil has not just been fuelling [US] military equipment and shaping [US] strategies; it has also been provoking the very wars in which these machines and tactics are employed" (Little 2009:49). Reportedly, "in 2009, the US military used 5.7 trillion gallons of oil, just under 16 million gallons a day. Its operations produced an estimated 73 million tons of greenhouse gases" (Angus and Butler 2011:174). A recent report indicates that "mobility fuel drives [Department of Defense (DoD)] energy spending: just 25% of DoD fuel is used for buildings and installations: the rest of the spending—about $12 billion per year—is for combat and combat-related systems" (Deloitte 2009:10). In 2006, jet fuel accounted for 58 percent and marine fuel for 13 percent of the total fuel consumed by the Department of Defense (Deloitte 2009:11).

An F-15 fighter jet burns about 1,580 gallons of oil per hour; a KC-10 aerial refueling tanker burns about 2,650 gallons per hour; and the B-52 bomber burns 3,266 gallons per hour (Little 2009:73). The outdated Abrams tank and the Bradley fighting vehicle both travel less than two miles for every gallon of petrol, and an armored Humvee gets about four miles per gallon.

The US military had some 30,000 vehicles situated in Iraq (Sanders 2009:47). The Department of Defense reported having a total inventory of 187,493 motor vehicles, including passenger cars, buses, and light trucks, 13 percent of them located overseas. The US Army operated over 4,000 combat vehicles and several hundred fixed wing airplanes. Top military officers had access to an airline called the Air Mobility Command, which included long-range C-17 Globemasters, C-5 Galaxies, C-141 Starlifters, and C-9 Nightingales (Sanders 2009:48). Generals and admirals had access "for their private use [to] seventy-one Lear jets, thirteen Gulfstream IIIs, and seventeen Cessna Citation luxury jets."

Military knowledge now includes the possibility of inducing climatic changes. The High Active Auroral Research Program (HAARP), part of the Star Defense Initiative, reportedly could potentially be used by the US military to modify the climate of enemy nations in order to destabilize their political economies (Chossudovsky 2004).

Ecological Footprints and Greenhouse Gas Emissions of the United States, China, and the United Arab Emirates

Despite the adverse impact of the Global Financial Crisis on the United States, it remains an economic dynamo, although very likely a declining one. Conversely, China has been on a rapid trajectory to becoming an emerging economic dynamo. These realities are illustrated in table 3.24. In interpreting these figures, one must bear in mind that whereas the United States has some 300 million inhabitants, China has some 1.3 billion

Table 3.24. Energy-Related and Economic Information: United States and China

		Actual (2004)	*Percentage of World Total*	*Projected (2030)*	*Percentage of World Total*
GDP (at market exchange rates, $US billion)	United States	10,704	30.6	22,494	29.3
	China	1,707	4.9	8,752	11.4
	World	34,937	100.0	76,850	100.0
Total energy use (quadrillion BTUs)	United States	100.7	22.5	131.2	18.7
	China	59.6	13.3	145.4	20.7
	World	464.7	100.0	701.5	100.0
Liquids/petroleum consumption (million barrels per day)	United States	20.7	25.1	26.6	22.6
	China	6.4	7.8	15.7	13.4
	World	82.5	100.0	117.6	100.0
Liquids/petroleum production (million barrels per day)	United States	8.2	9.7	10.2	8.7
	China	3.8	4.5	4.9	4.2
	World	84.3	100.0	117.7	100.0
Natural gas consumption (million cubic feet)	United States	22.4	22.5	26.1	16.0
	China	1.4	1.4	7.0	4.3
	World	99.6	100.0	163.2	100.0
Coal consumption (quadrillion BTUs)	United States	22.6	19.7	34.1	17.3
	China	41.1	35.9	95.2	47.8
	World	114.5	100.0	199.1	100.0
Nuclear energy consumption (billion kilowatt-hours)	United States	789	30.1	896	24.8
	China	48	1.8	329	9.0
	World	2,619	100.0	3,619	100.0
Carbon dioxide emissions (million metric tons)	United States	5,923	22.0	7,950	18.5
	China	4,707	17.5	11,239	26.2
	World	26,922	100.0	42,880	100.0

Source: Adapted from Klare (2008:246), based on data from U.S. Department of Energy, *International Energy Outlook 2007*.

inhabitants, more than four times as many. Given that the United States has about 4.4 percent of the world's population, the statistics indicate that it constitutes the culture of consumption par excellence.

In more recent figures comparing the United States and China, the International Energy Agency (2010:11) reports,

> In 2008, the United States alone generated 19% of world CO_2 emissions, despite a population of less than 5% of the global total. Conversely, China contributed a comparable share of the world's emissions (22%) while accounting for 20% of the world population.

The United States

The United States is often described as the richest country in the world. Indeed, compared with most countries, the United States is rich in many natural resources, although no longer in oil, as was the case in the past, or arable land, as is depicted in table 3.25.

Anthropologist John Bodley (2008:107) asserts that the "United States is without question the leading example of the *culture of consumption* as is illustrated by the fact that with less than five percent of the world's population, in 2000 it had 32 percent of the global GDP and consumed 22 percent of the world's energy consumption." Americans do not share equally in the country's culture of consumption as is evidenced by the amount

Table 3.25. Land Area Distribution Types for the United States and the World

	United States	*World*
Forests, woodlands	29	33
Pastures, rangelands	26	23
Agriculture	20	10
Desert	14	32
Built land	7	2

Source: Adapted from Vogt et al. 2010:23.

of poverty and homelessness in the richest country in the world. At the other end of the continuum, one finds avid overconsumers such as Ira Rennert, reportedly the 144th-richest person in the world. He owns 95 percent of Renco Group, a holding company with the US Magnesium Corporation and Doe Run Resources Corporation, the largest primary lead producer in the Western world, as its primary subsidiaries (Angus 2010:13). Some 10 years ago, Rennert had a new vacation house built in the Hamptons on Long Island (New York). Rennert's retreat reportedly is the largest contemporary residence in the United States, with 25 bedrooms, an equal number of bathrooms, 11 sitting rooms, 2 libraries, a servants' wing with 4 additional bedrooms, a power plant, 3 swimming pools, a $150,000 hot tub, and a 164-seat movie theater. He also owns an expensive house in Jerusalem and a luxury apartment in Manhattan and purchased twin $30 million apartments for his daughters. Rennert travels in a private Gulfstream jet.

As table 3.26 indicates, to sustain its economy, the United States relies heavily on fossil fuels. Its oil consumption increased by 69 percent from 1965 to 2008, with a significant reduction following the embargo of OPEC oil in the late 1970s and a minor reduction during the Global Financial Crisis that started in 2008

Table 3.26. Sources of US Energy, 2008

Source	*Use in Terawatts**	*Percentage*
Petroleum	1.24	37
Natural gas	0.80	24
Coal	0.74	23
Nuclear	0.26	9
Biomass (including ethanol)	0.14	4
Hydroelectric	0.07	2
Geothermal + solar + wind	0.03	1
Total	3.28	100
Total in GW = 3,280		

*A kilowatt is 1,000 watts of power; a megawatt is 1,000 kilowatts of power; a gigawatt (GW) is 1,000 megawatts of power; and a terawatt is 1,000 gigawatts of power.

Source: Adapted from Perlmutter and Rothstein (2011:40).

(Hughes 2010:213). To sustain its addiction to oil, the United States is the largest importer of oil in the world. Table 3.27 depicts the wide array of countries from which it obtains oil.

Oil played a critical role in the recent Global Financial Crisis starting in 2008 when its price started out at US$90 a barrel and peaked at almost $150. According to Tom Whipple (2010:252),

> By July 1, 2008, many industries that are dependent on oil, especially the airline and trucking industries, were desperate and in danger of being forced out of business. With the average price of gasoline above $4 a gallon in the United States (above $5 in California), car sales plummeted, leading to bankruptcy for much of the U.S. automobile industry and eventually massive government bailouts.

Table 3.27. Top Crude Oil and Petroleum Product Exporters to the United States, 2008

Country	*Exports to the United States (thousand barrels per day)*
Canada	2,499
Saudi Arabia	1,534
Mexico	1,305
Venezuela	1,192
Nigeria	991
Iraq	628
Algeria	550
Angola	514
Russia	466
Virgin Islands	321
Brazil	259
United Kingdom	237
Ecuador	221
Kuwait	211
Colombia	201
Other (80 countries)	1,821
Total	12,951

Source: US Energy Information Administration 2009.

In late 2008 oil prices had dipped to around $40 a barrel. The United States consumed 22 percent of the natural gas produced in 2008 (Hughes 2010:216). It also controls about 29 percent of the coal sources in the world (Hughes 2010:222). Although it continues to be a net exporter of coal, it has been importing coal from countries such as Colombia and Indonesia.

Coal constitutes the second-largest fossil fuel in the United States and dominates the country's electricity production, emitting about 10 percent of all global CO_2 emissions (Tomain 2010:54). In contrast to the US oil industry, which is dominated by a few multinational corporations, the US coal industry consists of "hundreds of coal producers operating about 2,000 mines" (Tomain 2010:54).

As the leading automobile culture in the world, much of the United States' greenhouse gas emissions come from motor vehicles, not only private cars but also huge trucks that haul many consumer commodities long distances across the country's vast landscape. Reportedly, transportation accounted for 29 percent of US greenhouse gas emissions and 5 percent of global greenhouse gas emissions in 2006 (US Department of Transportation 2010:2–5). The US Department of Transportation (2010:2–9) reports,

> GHG emissions from the U.S. transportation sector have been growing steadily—from 1990 to 2006, transportation GHG emissions increased 27 percent. The growth in U.S. transportation emissions accounted for almost one-half (47 percent) of the increase in total U.S. GHG emissions from the period. Emissions trends vary by transportation mode. Medium and heavy-duty truck GHG emissions increased 77 percent from 1990 to 2006, while light duty vehicles increased 24 percent, and aircraft 7 percent. On-road vehicles accounted for 96 percent of the increase in transportation emissions during that period; 55 percent from light-duty vehicles, 40 percent from medium and heavy-duty trucks, and one percent from other modes.

The International Energy Agency (2010:10) reports that the United States has the "highest level of passenger travel per capita in the world (more than 25,000 km [15,525 miles] per person per year)." The United States lags behind other developed countries and even China in requirements for greater fuel efficiency. Whereas in the United States, it has been 27 miles/gallon for cars and as low as 22 miles/gallon for SUVs, in the European Union it is 43 miles/gallon; in Japan, 46 miles/gallon; and in China, 36 miles/gallon (Perlmutter and Rothstein 2011:72).

In terms of electricity generation, in 2008 the United States obtained 49 percent of it from coal, 21 percent from natural gas, 20 percent from nuclear electric power, 6 percent from hydroelectric power, and 5 percent from other fuel sources (US Department of Energy 2008). The average new house in the United States increased in floor space by 40 percent between 1970 and 2004, thus requiring more energy to heat and cool and more appliances and other consumer items (Karlenzig 2010:297). Suburban and exurban sprawl have gone hand in hand with the trend toward larger dwelling units and contributed to growing dependence on private motor vehicles as the primary mode of transport. Residential and commercial buildings consume more than 70 percent of electricity produced in the country and constitute the "largest single source (43 percent) of carbon dioxide emissions" in the United States (Brown 2010:324). Reportedly, landfills containing many discarded consumer items constitute the "second-largest human-related source of methane in the United States, accounting for 23 percent of all methane emissions in 2007" (Sheehan and Spiegelman 2010:375).

Despite its tremendous wealth, the United States is also the most stratified of the core countries or advanced capitalist countries in the world and contains vast pockets of poverty. The Happy Planet Index developed by the New Economics Foundation seeks to determine how "happy" a country is by measuring the combination of its life expectancy and life satisfaction in relation to its ecological footprint. In 2009 the index ranked the United States 114th out of 143 countries (see McChesney and Foster 2010:5).

China

China "abandoned its decades-long attempt at autarkic development during the Mao Zedong years" and "elected to join the world market and build a modern industrial system on the basis of export-led growth, rather than through self-sufficiency and import substitution" (Jones 2005:109–10). China has become an economic powerhouse. According to Eugene Rosa and Thomas Dietz (2010:28), in 2003 it "consumed one-half of the world's cement production, one-third of its steel, nearly one-fourth of its copper, and nearly one-fifth of its aluminium," and it only lagged behind the United States as the world's leading oil importer. Indeed, China became a net importer of oil in 1993 to 1994, with Angola becoming its primary source of oil in 2005 (Vogt et al. 2010:89–95).

China's annual greenhouse gas emissions grew by 80 percent between 1990 and 2007 and surpassed those of the United States in 2007 (Christoff 2010). Xinjiang in the far western part of China reportedly contains 40 percent of the country's known coal reserves (Watts 2010:267). Inner Mongolia—particularly the city of Ordos—has evolved into China's largest coal-producing area (Watts 2010:370). China derives almost 70 percent of its energy from coal. As Jonathan Watts (2010:209) maintains, "cheap coal generates electricity for Beijing, Shanghai and Chongqing, fires the steel mills of Huaxi, power the production lines of Guangdong, and allows consumers in the West to buy Chinese goods at a knockdown price." To sate its thirst for oil, China has invested in oil companies in Angola, Indonesia, and Nigeria. In exchange for natural and manufactured resources from various African countries, including copper from Zambia, iron and steel from Zimbabwe, and even chocolate from the Ivory Coast, it is engaging in numerous aid projects as well as helping to "modernize African railroads and build highways, power stations, and dams" (Kaplan 2010:296). China relies on timber from countries such as Burma and Russia to sustain its world-leading wood-processing industry and role as a major exporter of finished wood products (Vogt et al. 2010:9).

None other than Pan Yue, the deputy director of China's State Environmental Protection Commission, has commented on how his country has mimicked the destructive patterns of developed capitalist countries:

> In 20 years, China has achieved economic results that took a century to attain in the West. But we have also concentrated a century's worth of environmental issues into those 20 years.

> While becoming the world leader in GDP growth and foreign investment, we have also become the world's number one consumer of coal, oil and steel—and the largest producer of CO and chemical oxygen demand (COD) emissions. (Quoted in Wall 2010a:92–93)

China's Five-Year Plan of 2006 includes plans to improve energy efficiency and to gradually shift from heavy reliance on fossil fuels to greater reliance on hydroelectricity, solar photovoltaic electricity, nuclear energy, and wind power. However, continuing heavy reliance on coal energy and the opening up of an ever-increasing number of coal-fired power plants have catapulted China ahead of the United States in terms of being the world's leading greenhouse gas emitter. While China still lags behind developed countries in both per capita resource consumption and per capita greenhouse gas emissions, as John Gulick (2011:27) observes, "this program could very well unleash into the atmosphere the extra increment of carbon dioxide that catalyzes runaway global warming, a catastrophe that would not only put paid to the chimera of Chinese hegemony, but would also devastate China's hundreds of millions of rural poor as severely as any other human collectivity in the world-system."

Tao Wang of the Tyndall Centre for Climate Change at the University of Sussex in the United Kingdom estimates that whereas emissions from exports from China in 2004 were 1,490 million tons of CO_2, emissions prevented due to imports were only 381 million tons of CO_2 (Wen 2010:132). Qin Gang, a spokesperson for China's Foreign Ministry, stated in June 2007 that

> China is now the factory of the world. Developed countries have transferred a lot of manufacturing to China. What many western consumers wear, live in, even eat, is made in China. (Quoted in Shiva 2008:35)

In a similar vein, Watts (2010:116) asserts that "governments and international corporations use China to sidestep the Kyoto

Protocol and other international treaties on environmental and labour standards."

The more affluent sectors of China's population have become important players in the global culture of consumption. China accounts for around 28 percent of the world's meat consumption (Myers and Kent 2004:20). Overall, meat consumption in China is around 50 kilograms per year as opposed to 122 kilograms per year for the average American, meaning that high meat consumption in China tends to be concentrated among the country's "new consumers" (Myers and Kent 2004:39). China has overtaken both Japan and the United States in terms of the number of mobile phones and is the third-largest personal computer market in the world, although one must bear in mind that China's population exceeds that of the United States by four times and that of Japan more than tenfold.

While power brokers and political pundits in the United States often express fear of the rising power of China, in reality this country may prove to be a "paper tiger" that is heading for an environmental crisis, including one induced by climate change, given that with about one-fifth of the world's population, it is "endowed with a mere 11% of the world's primary energy supply, and an even more meager seven percent of its arable land and freshwater reserves" (Gulick 2011:26). Climate change is already contributing to the desertification of parts of the country and diminished runoff from its mountains.

United Arab Emirates

The World Economic Forum ranks the United Arab Emirates (UAE) as the most competitive country in the Middle East (Krane 2009:138) Along with Qatar, Bahrain, and Kuwait, the UAE is among the leaders in terms of greenhouse gas emissions per capita in the world. It also consumes a great deal of water, much of it obtained from desalination plants, which are yet another source of greenhouse gas emissions (Vogt et al. 2010:84).

Abu Dhabi and Dubai, the two major cities in the United Arab Emirates, have quickly evolved into modern and highly energy-intensive, consumerist metropolitan hubs. The UAE is a federation of semi-independent emirates, with Abu Dhabi as the capital and Dubai as a tourist and shopping center and exemplar of capitalist decadence. In addition to millions of visitors each year, Dubai has a population more than 90 percent composed of temporary expatriate workers who also fuel the city's consumerist culture (Ali 2010:187). Abu Dhabi contains about 10 percent of the world's proven hydrocarbon deposits and accounts for over 90 percent of the UAE's oil exports (Davidson 2008:103). Conversely, Dubai is not as well endowed with oil but has managed to utilize its oil wealth to build more infrastructure in order to increase the emirate's portfolio (Davidson 2008:99–100). This consists of light manufacturing, several export processing (such as for agricultural products) "free zones," a luxury tourist industry, and a real estate market oriented to foreign investors. The Dubai Ports World Company is the fourth-largest port operator in the world (Davidson 2008:108). Dubai's tourist industry includes luxury hotels, horseracing, and golf courses (Davidson 2008:122). Dubai relies heavily upon air-conditioning to make its residents and visitors comfortable. Much of social life in Dubai centers around air-conditioned shopping malls, which compensate in part for the paucity of conventional public spaces, such as squares, parks, museums, and libraries, and provide relief from the "extreme heat and humidity that people must suffer for over half the year, often reaching 50 degrees Celsius in the summer months" (Ali 2010:9). Shopping, or the culture of consumption, constitutes a pivotal feature of "Dubai's position as a premier tourist destination in the region" (Ali 2010:43).

The UAE has been experiencing an annual increase in motor vehicles, many of which are luxury cars (Krane 2009:237). Of the latter, the UAE only ranks behind the much more populated United States in terms of sales of Humvees. To add injury to insult to the environment, the UAE reportedly experiences 116 annual deaths per 100,000 vehicles, 6 times the rate for the United States and 10 times that of the United Kingdom. Emirates Airlines based in Dubai undoubtedly produces an incredible amount of greenhouse gas emissions as its planes provide flights to cities in the Middle East, South Asia, Europe, East Asia, Africa, the United States, Australia, and New Zealand. As Syed

Ali (2010:21) observes, Emirates "benefits to a large degree from geography, lying as it does at the intersection of Europe, Africa, and Asia."

Conclusion

Global capitalism, which relies heavily upon fossil fuels, has played a significant role in the emission of greenhouse gases into the atmosphere, which in turn has contributed to climate change. The anthropogenic sources of climate change include a growing global population clamoring for both basic and luxury goods; the growing proliferation of motor vehicles; a growing number of airplane flights around the world; the construction of factories, offices, shopping malls, and dwelling units; the overheating and overcooling of these facilities; industrial agriculture; deforestation; the production of consumer products (many of which are not necessary for subsistence); militarism; and various other operations. Given that developing countries such as China and India have embarked upon rapid programs of industrialization and stepped up their patterns of consumption, it can be expected that greenhouse gas emissions from developing countries, with their teeming populations, will soon exceed those from the developed world.

As a result of the Global Financial Crisis, air travel and house and car sales declined in the United States and many European countries in the second half of 2008 (Alier 2009:1105). World GDP decreased by 1 or 2 percent in 2009, with economic degrowth even greater in the United States, the European Union, and Japan (Alier 2009:1108). The World Bank (n.d.), in an article titled "Growth Will Slow but Remain Robust," reported that in the United States, in the wake of "relatively weak weather-influenced first quarter GDP results [in 2011], and some flagging in the pace of the recovery in the second quarter, GDP growth is expected to pick up in the second half of the year, with whole year gains of 2.6 percent in 2011 and 2.9 percent in 2012, and with growth easing to 2.7 percent by 2013." Despite the World Bank's cautious optimism, Immanuel Wallerstein (2011b:2) asserts that the United States and the global economy have "moved into an era of acute, constant, and rapid fluctuations—in exchange rates of currency, in rates of employment, in geopolitical alliances, in ideological definitions of the situation." Despite an economic downturn, capitalism still has tremendous potential to reverse such trends and resume "business as usual." The treadmill of production and consumption that is an inherent component of global capitalism, with its emphasis on profit making, need for continual economic expansion, and heavy reliance on fossil fuels, makes it an environmentally unsustainable and destructive system as well as a generator of climate change. Not only eco-Marxists or eco-socialists have arrived at this conclusion but also various more conventional scholars, such as James Gustave Speth, a longtime environmental insider who served as chairperson of the Council on Environmental Quality in Jimmy Carter's administration, a senior advisor on Bill Clinton's transition team, and the administrator of the United Nations Development Programme between 1993 and 1999. Speth (2008) argues that existing capitalism is highly destructive of the environment, to the point that it threatens life on the planet in its present form. He does not view socialism or even eco-socialism as an alternative to contemporary capitalism but instead advocates a variant of green social democracy that would rely on markets but be committed to a "New Sustainability World." However, as I argue later in this book, achieving sustainable or green capitalism is a highly dubious proposition given capitalism's predilection for continual growth or expansion.

4

The Inadequacies of Existing Climate Regimes for Mitigating Climate Change

The grim reality of climate change has provoked the spilling of much ink and the emptying of many printer cartridges in discussing how best to adapt to it over the next decades and how to mitigate it over the long run. While it is inevitable that, over the short run, human societies will have to adapt to climate change, the more crucial issue is mitigation—that is, transcending climate change to ensure the survival of humanity as well as to maintain biodiversity. Indeed, a mere focus on adaptation poses the danger of political complacency, fatalism, and even cynicism. Numerous strategies have been proposed to address climate change, most of which seek more or less to work within the parameters of the existing capitalist world system. This chapter discusses two broad categories of climate change strategies: *climate regimes* and *green capitalism*. Various scholars and policy advisors have argued that the problem of climate change will require some kind of internationally coordinated response. Regimes refer to rules and decision-making processes in which nation-states agree to defer their sovereignty to a large international body. Indeed, since the late 1980s, climate change regimes have emerged at the international, regional, national, provincial, state, and even local levels. The vast majority of climate regimes function within the parameters of green capitalism, a notion that capitalism, by adopting various technological innovations, energy efficiency, recycling, and other practices, can be environmentally sustainable. While historically corporations have been resistant to the assertion on the part of environmental activists that many of their practices are environmentally destructive and contribute not only to environmental damage as well as climate change, a growing number of corporations have begun to assert that they can engage in both sustainable development and reduce greenhouse gas (GHG) emissions by engaging in a process of ecological modernization.

Climate Regimes

Joseph Camilleri and Jim Falk (2010:292) assert that "global warming has paradigmatically changed the scope of human governance." Climate regimes, however, include governments (national, state or provincial, and local), business and industrial nongovernmental organizations (BINGOs), and environmental nongovernmental organizations (ENGOs). BINGOs can be divided into (1) "green business" bodies such as the US-based Business Council for Sustainable Energy, the European Business Council for a Sustainable Future, and the World Business Council for Sustainable Development; and (2) "grey business" bodies, which downplay or deny the reality of anthropogenic climate change, such as the now defunct Global Climate Coalition. Many ENGOs have become part and parcel of the climate movement, both at the international and national levels, a topic explored further in chapter 8. At the national level, climate regime politics tends to be shaped by *lead states,* such as most EU nations and more recently Bolivia, and *veto states,* such as the United States, Australia (at least until late 2007, when the government of Kevin Rudd ratified the Kyoto Protocol), China, India, and various oil-producing countries (Carter 2007:238).

United Nations–Related Bodies

The first World Climate Conference recognized in 1979 that climate change constitutes a serious threat to humanity and the planet and inaugurated a series of meetings on how to reduce greenhouse gas emissions (Hossay 2006:13). In 1987 the UN Commission on Sustainable Development released a document entitled *Our Common Future* acknowledging that climate change constitutes a major environmental dilemma for the future of humanity (Skjærseth and Skodvin 2009:128). The following year the UN General Assembly passed a resolution stating that climate change constitutes a "common concern for mankind" (Yamin and DePledge 2004:22). In 1988 the United States sponsored a technical workshop in Villach, Austria, that asserted that greenhouse gases were very likely contributing to climate change (Davis 2007:186). In 1988 Canada sponsored the Toronto Conference on the greenhouse effect, which recommended a 20 percent reduction in CO_2 emissions by 2005, the creation of an international treaty addressing climate change, and a fund through which developed countries would assist developing countries in reducing their greenhouse gas emissions. David Howard Davis (2007:185–86) asserts, "The Toronto Conference marked the transition of the global warming issue from a scientific forum to a government forum." The UN Commission on Sustainable Development met again in Villach in 1990 and later in Bellagio, Italy.

The Intergovernmental Panel on Climate Change

Eminent Swedish climatologist Bert Bolin persuaded the United Nations to form the Intergovernmental Panel on Climate Change (IPCC) (Lovelock 2009:3). It came out of an initiative on the part of the World Meteorological Organization and support of the United Nations Environment Program. As Camilleri and Falk (2010:296) argue, the IPCC symbolizes the "extent to which global warming [has] raised the role of science to a position of pre-eminent engagement with, and influence in, policy development and advocacy." The IPCC played a pivotal role at the UN Conference on the Environment and Development, better known as the Earth Summit, in Rio de Janeiro in 1992. The IPCC is governed by a joint task force plus three working groups, all cochaired by one representative each from a developed country and a developing country. Working Group I assesses the scientific dimensions of the climate system and climate change. Working Group II assesses the vulnerability of human societies and natural systems to climate change, the purported negative and positive consequences, and the possible strategies for adapting to them. Working Group III assesses the options for limiting greenhouse gas emissions and mitigating climate change, as well as its economic implications. In essence, the IPCC provides governments and the global community with scientific, technical, and socioeconomic information pertinent to evaluating the risks of climate change and responding to it. It does not conduct independent scientific research but rather compiles key research and seeks to produce consensus documents. The IPCC published assessment reports in 1991, 1997, 2001, and 2007 and is slated to publish its next report in 2014. The 2007 IPCC assessment involved approximately 400 experts from some 120 countries who were involved in drafting, revising, and finalizing the three major reports (Maslin 2009:14). The participants in the entire IPCC reporting process consist of over 2,500 scientific expert reviewers, over 800 contributing authors, and over 450 lead authors (Craven 2009:120).

With each report, the IPCC has been more assertive in making the claim that human-related activities constitute the main reason for climate change in recent times. The Second Assessment maintained that the "balance of evidence" suggested a "discernible human influence on global climate" (Camilleri and Falk 2010:276). The Third Assessment stated that "most of the observed warming over the last 50 years is *likely* to have been due to the increase in greenhouse gas concentrations." The IPCC more boldly asserted in its Fourth Assessment that "very likely"—in other words, more than 90 percent probability—anthropogenic greenhouse gas emissions have been responsible for most of the increase in global warming. The IPCC along with Al Gore received a Nobel Prize in December 2007.

The IPCC is very much a contradictory body in that, on the one hand, it does provide strong evidence that climate change constitutes a serious problem and needs to be addressed; on the other hand, as Larry Lohmann (2006:36) so astutely observes, it "has helped shape climate problems and solutions in such ways that make them more acceptable to powerful governments and corporations." He goes onto argue that

> much of the IPCC's work had the effect of making climate change seem potentially manageable by private and public sector institutions including oil companies and the World Bank, and by means of neoliberal approaches generally. It became "politically incorrect" to enquire whether radical social change might be necessary to reduce greenhouse gas concentrations to a safer level. What was needed, it was implied, was to unleash the productive forces of private sector companies in the service of climate stability. (Lohmann 2006:40)

The May 1990 IPCC meeting in the United Kingdom focusing on the final draft of the first Scientific Assessment Report included 11 scientists employed by fossil fuel industries (Godrej 2006:92). One of them, Brian Flannery, an Exxon employee and a representative of the International Petroleum Industries' Environmental Conservation Association, argued that there are uncertainties about how carbon affects the climate system and questioned the validity of IPCC climate models. Fortunately, most of the other IPCC scientists maintained that such uncertainties should not affect the ultimate goal of stabilizing the amount of atmospheric CO_2 and rejected his dismissal of climate models. Nevertheless, this incident illustrates that the fossil fuel industry has been able to penetrate the bowels of the leading climate system assessment body in the world. An IPCC working group met with government representatives in August 1990 and warned them of the dire consequences if greenhouse gases continue to rise. According to Dinyar Godrej (2006:97),

> But all that the policy responses working group had come up with, after 18 months of deliberation, was a toothless list of potential technologies which could help, in principle, with the limitation of greenhouse gases. This third working group was chaired by the United States.

Efforts to write the executive summary of the IPCC Second Assessment Report stumbled over the phrase "changes point towards a human influence on climate," which was watered down to "the balance of evidence suggests that there is a discernable human influence on global climate" (quoted in Godrej 2006:99–100). Delegates representing the Saudi and Kuwaiti governments and advised by Washington attorney Don Pearlman, a climate change denier, questioned the working group frequently. As a consequence of this tedious process, the executive summary "had shrunk to a quarter of its original size mainly due to the constant stalling by vested interests" (Godrej 2006:100).

Nevertheless, the 1996 IPCC report did bring the issue of the urgency of reducing greenhouse gas emissions to the Conference of the Parties (COP) agendas. Stephen Schneider in *The Patient from Hell* recalls his part in the development of the IPCC Working Group II report of 2001 (see Lovelock 2009:8). He describes how good science presented at the session was manipulated until it satisfied all of the national representatives present. Robert T. Watson, the World Bank scientist who served for a while as the head of the IPCC, eventually lost his position in 2002, reportedly as a result of offending ExxonMobil, which had contributed handsomely to George W. Bush's reelection fund, with the US Department of State opposing Watson's reelection (Lohmann 2006:41; Godrej 2006:116). Rajendra Pachuari, the new IPCC chairperson, proved a more formidable force in terms of the serious nature of climate change.

The 2007 IPCC report's estimates on sea level rise by 2100 "may have trumped the science" (Ward 2010:26). The IPCC delineated the following scenarios going from the lowest to the highest rise (see Ward 2010:28):

B1 scenario: 0.18–0.28m rise in sea level
A1T scenario: 0.20–0.45m
B2 scenario: 0.20–0.43m
A1B scenario: 0.21–0.48m
A2 scenario: 0.23–0.51m
A1Fl scenario: 0.26–0.59m

Whereas the IPCC in its Third Assessment Report had projected a maximum sea level rise of 0.70 meters by 2100, it had dropped this maximum to 0.59 meters in its Fourth Assessment Report. According to Ward (2010:29), "almost immediately after publication of the IPCC 4, new studies of *past* sea level rises . . . put the IPCC estimates into question." James Hansen maintains that sea level rise could be over three feet (or a meter), if not higher, by 2100 and "sees the current climate change situation as causing many scientists to be more worried about crying wolf than 'fiddling while Rome burns'" (Ward 2010:36). According to Chivers (2007:65–66), the 2007 IPCC report "was stripped of many 'undesirable' passages by politicians before it could be published, including warnings about the likely impacts of climate change on North America and references to positive feedback loops and the risk of runaway climate change."

One of the glaring shortcomings of the IPCC is the underrepresentation of authors from developing countries. With respect to the IPCC Fourth Assessment Report, Working Group I included 125 authors from Organization for Economic Cooperation and Development (OECD) countries and 44 from non-OECD countries; Working Group II, 146 authors from OECD countries and 82 from non-OECD countries; and Working Group III, 122 authors from OECD countries, 72 from non-OECD countries (Barnett and Campbell 2010:62). Overall, 393 authors were from OECD countries, and 198 were from non-OECD countries. Indeed, "some developing countries . . . do not send delegates for meetings of the panel" (Barnett and Campbell 2010:61).

Despite its members' best intentions to mitigate climate change, the IPCC as a scientific assessment and policy advisory body has not been able to contain the increase in greenhouse gas emissions associated with the capitalist treadmill of production and consumption. As Luke (2008:141–42) observes,

> The social critique of climatology is real, but can it ever get past accepting sustainable degradation? Climatologists admit there indeed is a crisis, and then they seek to respond in a proactive, profitable, and powerful fashion. Yet, does the work of the IPCC only mask negative outcomes, maintain some environmental viability, and create zones of control where degradation is at best lessened, but greenhouse gassing is never stopped? The growing number of scientific studies heighten awareness of climate change, yet it is rarely stemmed. The existing inequality of commodity production and consumption

> spills over into new inequalities in commodity by-production and consumer choicelessness, because technoscience is left only to scrupulously document additional biospheric losses. However, it cannot easily change how loss is incurred.

The Framework Convention on Climate Change and the Kyoto Protocol

Camilleri and Falk (2009:260–64) delineate three phases in what they term the *Kyoto process*: Phase 1, which started with the Toronto Conference in 1988 and culminated in May 1994 with the creation of the Framework Convention on Climate Change (FCCC); Phase 2 (1995–1997), which focused on the implementation of FCCC; and (3) Phase 3 (1997–2004), which resulted in the ratification of the Kyoto Protocol.

The first IPCC report in 1991 paved the way for the United Nations Conference on Environment, also known as the Earth Summit, held in Rio de Janeiro in 1992, an event attended by some 10,000 delegates from 178 countries. The Earth Summit in turn resulted in the formation of the Framework Convention on Climate Change, which took force in 1994. This body was modeled upon the 1996 Montreal Protocol, an international treaty that contributed significantly to the reduction of ozone-depleting chlorofluorocarbons (CFCs) used in various products, such as air conditioners and refrigerators. The Montreal Protocol is widely regarded as one of the most successful international environmental regimes in that it resulted in a reduction of CFCs from 1.2 million tons in 1986 to 164,000 tons in 1997 (Carter 2007). The UN FCCC aims to achieve "stabilisation of greenhouse gas concentrations in the atmosphere at a level that would prevent dangerous anthropogenic interference with the climate system."

The UN FCCC has since 1995 held a series of Conference of the Parties meetings more or less on an annual basis (table 4.1). While the developed countries tend to dominate the meetings because they have more resources, in theory every signatory country to the UN FCCC has an equal right to participate. The convention divided countries into two categories: (1) Annex

Table 4.1. FCCC Conferences of the Parties

Number	*Location*	*Year*
1	Berlin	1995
2	Geneva	1996
3	Kyoto (protocol adopted)	1997
4	Buenos Aires	1998
5	Bonn	1999
6	Hague (Part 1)	2001
6	Bonn (Part 2)	2001
7	Marrakech	2001
8	New Delhi	2002
9	Milan	2003
10	Buenos Aires	2005
11	Montreal	2005
12	Nairobi	2006
13	Bali	2007
14	Poznan	2008
15	Copenhagen	2009
16	Cancún	2010
17	South Africa	2011

I countries, which are developed and are OECD members that have historically emitted the most greenhouse gases, and (2) non–Annex I countries, which are primarily developing and were not OECD members in 1990. In reality, various non–Annex I countries, such as China, South Korea, Mexico, Brazil, and South Africa, have been undergoing a considerable amount of industrialization in recent decades. After Russia ratified the Kyoto Protocol on November 18, 2004, the agreement took force on February 16, 2005, having been signed by 141 countries representing 85 percent of the world's population.

The Conference of the Parties serves as the governing body of the convention. COP1 occurred in Berlin in 1995. COP3, which assembled in December 1997, adopted the Kyoto Protocol and reportedly was attended by over 10,000 people, including 1,500 delegates from 160 countries, 3,500 observers, and 4,000 media people. It resulted in a scheme generally known as the Kyoto

Protocol, inspired by the Montreal Protocol, which dealt with ozone emissions contributing to the creation of an ozone hole in the atmosphere. According to Lohmann (2006:46), in contrast to the challenges of effective climate change mitigation, the Montreal Protocol did not pose a serious threat to multinational corporations, in that

> only a few factories were involved. It was relatively easy to set a target and find substitutes for some ozone-depleting substances or phase them out. With the eventual backing of industry itself and the help of a few transition-aiding payments to Southern nations, nearly all nations wound up complying with the agreement.

The success of the Montreal Protocol resulted in part from the availability and rapid introduction of non-ozone-depleting alternatives, one of which is hydroflurocarbon, used in refrigeration and unfortunately a potent greenhouse gas. The Kyoto Protocol has not succeeded in finding readily available alternatives to fossil fuels.

The Kyoto conference entailed much horse-trading among the delegates, with strong opposition to significant emissions cuts coming from the oil-producing nations, such as Saudi Arabia and Kuwait; Iceland, with its large natural gas reserves; Norway, with its North Sea oil reserves; and the United States, Canada, and Australia, all big emitters (Godrej 2006:107). Conversely, the Alliance of Small Island States (AOSIS) argued for cuts of 20 percent below 1990 levels. During the negotiations, Brazil submitted a proposal that sought to apportion greenhouse gas emissions targets according to each country's historical responsibility for global temperature increases (La Rovere et al. 2002:157). The Brazilian government and national scientific community experts had developed this scheme in 1996 and 1997; developed countries rejected the proposal.

The Kyoto Protocol seeks to reduce emissions of six gases—namely, carbon dioxide, methane, nitrous oxide, sulfur hexafluoride, HFCs, and PFCs. It is organized on the basis of an emissions trading scheme (ETS). According to Chivers (2009:133), "Under huge domestic pressure from fossil-fuel lobbyists (including a $23-million anti-Kyoto television campaign) the US delegation, led by Al Gore, only [signed] up on the condition that emissions trading [was] included in the deal."

Articles of the Kyoto Protocol pertinent to an ETS include

- Article 3.1, which specifies that countries can meet objectives jointly, such as in the case of the European Union
- Article 3.13, which permits countries to bank unused emissions during the period from 2008 to 2012
- Article 6, under which emissions credits can be earned using emissions reductions Annex B countries subject to binding targets and authorized to exchange these credits
- Article 12, which allows Annex I countries to earn additional emissions reductions in non–Annex I countries by participating in the Clean Development Mechanism (CDM) scheme

The Kyoto Protocol emissions accounting system does not including shipping and aviation. It also adopts a geographical approach to emissions responsibility, which means that emissions generated from production in a particular country are attributed to that country. It does not factor in that developed countries often consume the products manufactured in developing countries, such as China. In large part, the emissions trading paradigm that underlies the Kyoto Protocol was an American-inspired policy applied in the 1970s in the United States in dealing with lead, nitrogen oxides, sulfur dioxide, and other pollutants and that was "successfully pressed on the UN by the US government [including Al Gore when he served as vice president in the Bill Clinton administration], advised by US economists, US NGOs and US business, in the 1990s" (Lohmann 2006:48). Ward (2010:64) contends that the "United States and other powerful nations hamstrung the agreement from the start, by allowing it to exempt fourteen countries considered 'economies in transition' including China and India, which both ratified the treaty merely as a public relations move." Furthermore, Gore argued that the US Congress would not ratify the Kyoto Protocol if a carbon tax rather than an emissions trading scheme were adopted. Ironically, in the end, the United States did not ratify the Kyoto Protocol and has not yet done so, although Australia finally did so when the government of John Howard was replaced by the Rudd government in late 2007. Unfortunately, largely due to pressure from the Clinton administration, the Kyoto Protocol contains at least three ETSs—namely, a cap-and-trade program that constitutes its mainstay, as well as the Clean Development Mechanism and the Joint Implementation, which I discuss later in this book (Driesen 2010:136).

The failure of the United States to ratify the Kyoto Protocol ultimately meant that it could not take effect until Russia ratified the agreement on February 16, 2005. According to Arnaud Brohé, Nick Eyre, and Nicholas Howarth (2009:64):

> Russian participation was essential (following the refusal to ratify by the US) as a prerequisite for the entry into force of the Protocol is that ratifying parties cover at least 55 percent of the total CO_2 emissions of all Annex I Parties to the Convention. In practice, Russia's target is relatively easy to meet as its emissions have declined substantially through deindustrialization since 1990.

Russia is in a position to sell its "hot air" credits when it chooses, enabling developed countries to purchase the credits and thereby meet their Kyoto targets without actually reducing emissions.

The Kyoto Protocol requires Annex I countries to commit during the 2008–2012 period to reducing their greenhouse gas emissions in most cases, while allowing some countries to increase emissions during the commitment period. Table 4.2 depicts the commitments to cut emissions of various countries under the Kyoto Protocol. Despite the fact that Australia has long been a major greenhouse gas emitter per capita, it managed to obtain credit for having reduced its rate of land clearance due to increasing public concern about its environmental impact.

Table 4.2. Commitments under the Kyoto Protocol

Percentage of the Reference Level					
Australia	108	Greece	92	Norway	101
Austria	92	Hungary	94	Poland	94
Belgium	92	Iceland	110	Portugal	92
Bulgaria	92	Ireland	92	Romania	92
Canada	94	Italy	92	Russia	100
Croatia	95	Japan	94	Slovakia	92
Czech Republic	92	Latvia	92	Slovenia	92
Denmark	92	Liechtenstein	92	Spain	92
Estonia	92	Lithuania	92	Sweden	92
European Union	92	Luxembourg	92	Switzerland	92
Finland	92	Monaco	92	Ukraine	100
France	92	Netherlands	92	United Kingdom	92
Germany	92	New Zealand	100	United States	93

Given that land clearance had been a major source of Australian greenhouse gas emissions, the country was credited for its efforts with permission to increase its emissions. The Kyoto Protocol exempts developing nations, including China, India, and Brazil, from setting emissions reductions targets. It was intended to initiate a series of treaties that would eventually be binding at least upon countries that have ratified it.

Despite the implementation of the Kyoto Protocol, as table 4.3 indicates, its signatories have been quite variable in terms of the extent to which they have reduced their CO_2 emissions, bearing in mind that some signatories were permitted an increase in their CO_2 emissions. Germany is often regarded as an international leader in climate change mitigation efforts because it agreed to cut back CO_2 emissions by 21 percent from 1990 to 2012 under the provisions of the Kyoto Protocol. Indeed, as table 4.3 indicates, it appears very likely that Germany will meet its 2012 target. However, much of Germany's initial advance toward this target resulted from the closure of large sectors of the highly polluting East German infrastructure. Germany has

Table 4.3. Total Anthropogenic CO_2 Emissions of Selected Signatories, Including Emissions/Removals from Land Use, Land-Use Change, and Forestry, 1990 and 2008

	Gg CO_2 Equivalent Change from 1990 to 2008		
Country	*1990*	*2008*	*Percentage*
Australia	317,190	464,036	46.3
Canada	398,214	554,227	39.2
Denmark	53,235	53,498	0.5
European Union	4,048,199	3,670,898	−9.3
Finland	40,516	22,585	−44.3
Germany	1,016,438	862,488	−15.1
Iceland	4,228	5,278	24.8
Ireland	32,609	45,878	40.7
Japan	1,079,972	1,135,599	5.2
Poland	441,832	282,375	−36.1
Russian Federation	2,612,573	1,070,391	−59.0
Spain	189,214	285,721	51.0
Sweden	25,491	35,606	39.7
United Kingdom	594,071	534,593	−10.0
United States	4,191,418	4,980,855	18.8

Source: FCCC, national greenhouse gas inventory data for the period 1990–2008, 21.

been actively developing wind and solar energy sources with the objective of having them account for 50 percent of all energy sources by 2050 (DiMento and Doughman 2007:112).

The Clinton administration signed the Kyoto Protocol in 1998 but encountered heavy opposition from the US Senate, which opposed the decision by a vote of 95–0 on the grounds that the protocol would harm the US economy and did not include developing countries. Numerous US corporate interests engaged in intense lobbying to oppose ratification over the course of 1998. Australia initially supported the protocol, but the Howard government, which assumed power in 1996, backed off on the grounds that the developing countries had not signed and that the provisions of the protocol would interfere with the growth of the Australian economy.

COP6 held in The Hague in November 2000 was largely a dismal failure (Hamilton 2007:89). The more reluctant signatories of the protocol formed an umbrella group consisting of the United States, Canada, Japan, Australia, and a few small countries, which sought loopholes in the original agreement. The United States insisted that it be permitted to set off its emissions against its carbon sinks, particularly in the form of vast forests (Carter 2007). While the George W. Bush/Al Gore electoral fiasco was playing itself out in the United States, Frank Loy, one of the US negotiators, stated that he would do nothing that would "'jeopardize the American lifestyle" (Godrej 2006:113). When Bush became president in January 2001, he unequivocally opposed ratification of the protocol. When the new Australian Labor Party government under the helm of Prime Minister Kevin Rudd ratified the protocol in December 2007, the United States became the only remaining Annex I country not to have ratified the agreement and has maintained this dubious status even under the watch of President Barack Obama.

COP11 in Montreal in November and December 2005 constituted the first convention meeting after the Kyoto Protocol came into force. COP13 held in Bali in December 2007 set a road map with the following items: (1) the scheduling of COP14 in Poznan, Poland, in December 2008; (2) the scheduling of COP15 in Copenhagen in December 2009 with the objective of concluding convention negotiations for a post-2012 framework; and (3) the ratification of a new climate agreement in 2012.

At the Bali conference in 2007, the International Emissions Trading Association constituted the largest NGO delegation (*Green Left Weekly*, December 12, 2007). It consisted of 336 representatives, including lawyers, financiers, emissions traders, consultants, certifiers, and emissions trading experts from corporations such as Shell and accounted for 7.5 percent of the 4,483 NGO delegates at the conference. In contrast, the World Wildlife Fund accounted for 2 percent of the NGO delegates and Greenpeace, 1.6 percent. Although George W. Bush had admitted before the Bali conference that climate change was real and largely anthropogenic, his delegation led by Undersecretary of State Paula Dobriansky made a concerted effort to act as an obstructive force. However, "on the conference's final day, delegates from most of the 190 nations booed Dobriansky loud and long—a sustained global jeer that won American acquiescence on a small point and let everyone sign the Bali Road Map [which was designed to lead to a new treaty in Copenhagen in December 2009], such as it was, and go home" (Pooley 2010:5).

In 2008, 192 countries and the European Union were parties to the convention. According to Brian Tokar (2009b), "Corporate representatives have been hovering like vultures over UN climate meetings, seeking to define the terms of what they hope will be a rapidly expanding market in tradable carbon allowances, and the World Bank is jockeying to control the funds to curtail deforestation." The Bali conference officially implemented the Reducing Emissions from Deforestation and Forest Degradation (REDD) scheme as a climate change mitigation strategy. A major dilemma associated with REDD is that "indigenous people face the risk that governments, companies, and conservation NGOs will 'zone' forests, thereby creating protected areas, biological corridors, forest reserves and sustainable forest management zones in order to receive REDD payments while excluding or disadvantaging indigenous and traditional communities" (Lyster 2010:122–23). The US-based NGO Environmental Defense Fund (now called Environmental Defense) advocated as early as 1991 emissions trading as a mean for allegedly protecting the rainforest (Gilbertson and Reyes 2011:22).

At the Copenhagen conference, the Ad Hoc Working Group on Long-Term Cooperative Action was reportedly prepared to endorse REDD-plus, which would recognize the "rights of indigenous peoples and members of local communities." REDD-plus would entail three basic principles: "participation in political decision-making, equitable distribution of forest benefits, and recognition of forest people's particular identities" (Sikor et al. 2010:423).

The Kyoto Protocol permits Annex I countries to trade their allocated carbon emissions among themselves as well as to invest in carbon mitigation projects in developing countries under the provisions of the Clean Development Mechanism, which has been described as a "public-private partnership" (Newell and Paterson 2010:29). In essence, emissions trading schemes allow polluters to continue polluting by purchasing credits. Facilities that exceed their allocation can purchase credits and thus avoid actually reducing greenhouse gas emissions. Whereas 2 percent of CDMs consist of renewable energy projects, 72 percent of them are based on carbon capture, and 21 percent are based on biomass (Shiva 2008:18). India has the greatest number of CDM projects, 82 as of 2006, with another 350 being planned (Shiva 2008:22). Giddens (2009:190) queries the effectiveness of CDM in introducing renewable energy projects into developing countries:

> Marginal projects dominate, such as the containment of industrial gases by bolting on filters to already existing pipes. It has been said that perhaps half the reductions claimed are the result of "accounting tricks" and are empty of content. In one case, the projects in a specific country were all concerned with emissions of HFC-23, a by-product of the manufacture of refrigerants.

As noted earlier, the emissions targets set by the Kyoto Protocol are very modest, calling for an average of 5.2 percent below 1990 levels by 2012. Now, into 2012, various preliminary figures indicate that most signatories are nowhere close to meeting their designated targets. Even if they were to do so, the commitment of Annex I countries to an overall 5 percent reduction by 2012 is only a fraction of the emissions reduction needed to achieve a safe climate. COP15, which assembled in Copenhagen in December 2009, theoretically sought to finalize a new international climate agreement that would serve as a successor to the Kyoto Protocol, which expires at the end of 2012. It was the largest international climate conference ever, involving nearly 27,000 people, including 10,500 delegates from 190 countries and more than 120 heads of state. The European Union proposed reducing emissions by 30 percent by 2020 and suggested a mechanism of funding the energy transition in developing countries, but it did not receive support from "climate laggards" such as the United States and Saudi Arabia. Hugo Chávez, the president of Venezuela, may have been the most colorful presence at the conference when he blamed global capitalism as the source of climate change and stated, "If climate change was a bank, it would have already been saved!" (quoted in Lowe 2010:258).

The Copenhagen conference failed to produce a new treaty with binding mechanisms to keep the global mean temperature below 2°C, a temperature target based on the 2007 IPCC report. The Copenhagen Accord constituted a desperate effort to mask the failure of Copenhagen and was constructed during the last 24 hours of the conference by a small group of government leaders. The Annex I group and most developing countries supported the accord. Conversely, the accord was opposed by seven countries, initially Tuvalu, Nicaragua, Bolivia, Cuba, Venezuela, Sudan, and later Pakistan (Dimitrov 2010:20). The accord acknowledges the 2°C temperature target and the need to make deep cuts in global greenhouse gas emissions to stabilize the Earth's climate but does not stipulate any binding timetables for emissions reductions and lacks any compliance mechanisms. Carl Death (2011:10) presents a sobering overview of the flaws of the conference in his observation that "the emerging prominence of the so-called BASIC group (Brazil, South Africa, India, and China) in Copenhagen, together with the United States, points towards the probable shape of the new coalition driving global climate policy, at the expense of UN multilateralism."

As of December 2011, 141 countries, which account for about 87 percent of global greenhouse gas emissions, had pledged under the provisions of the Copenhagen Accord to limit their emissions by 2020. The United States provided a limit of 17 percent below 2005 levels, whereas the European Union offered a range of 20 to 30 percent reductions. J. Rogelj et al. (2010) conducted an analysis that considered "estimates from the pledges submitted to the Copenhagen Accord and, for countries that didn't submit targets, from previous announcements" and assumed business-as-usual growth for the remaining countries that did not announce targets and concluded that "temperatures would even exceed three degrees Celsius warming by 2100." The accord includes a provision under which developed countries will provide $30 billion in 2010, increasing to $100 billion per year from public and private sources by 2020, to help developing countries mitigate emissions, preserve forests, and adapt to climate change.

Peter Christoff (2010:651) provides the following sobering assessment of the failure of the Copenhagen conference to deliver a binding agreement to reduce greenhouse gas emissions:

> Current pledges under the Accord cannot deliver its 2°C goal, but leave the world in peril of global warming of over 4°C above pre-industrial levels by 2100. The very limited window of opportunity for significant action to avoid catastrophic global warming may thus be missed.

While the vast majority of the delegates at COP15 viewed climate change mitigation within the parameters of global capitalism, some, such as Hugo Chávez from Venezuela and Evo Morales from Bolivia, did not. Morales stated at the conference that

> living better is to exploit human beings. It's plundering natural resources. It's egoism and individualism. Therefore, in those promises of capitalism, there is no solidarity or complementarity. There's no reciprocity. So that's why we're trying to think about other ways of living lives and living well, not living better. Living better is always at someone else's expense. Living better is at the expense of destroying the environment. (Quoted in Magdoff and Foster 2010:28)

In 2010 at COP16 in Cancún, all parties, except Bolivia, agreed to the Cancún Agreement, which permits the Kyoto Protocol to continue to operate while at the same time weakening the obligations of developed countries to cut their greenhouse gas emissions by expanding their reliance on carbon trading and financial and technological transfers to developing countries. COP15 gave formal recognition to the need to stabilize global temperatures below 2°C and to consider strengthening this limit to 1.5°C. It created a new Climate Fund, and the World Bank, a major investor in fossil fuel projects, was nominated to administer the proposed $100 billion. The Cancún conference also created the space for the inclusion of carbon capture and storage (CCS) as a CDM scheme (Friends of the Earth Europe 2010a:7). Finally, it established a new international program that would further REDD activities in developing countries. In essence, the Cancún Agreement served as yet another instance of locking market-based mechanisms into climate change mitigation policy.

COP17 took place in Durban, South Africa, from November 28 to December 11, 2011. Whereas the European Union, the poorest of the developing countries, and small island states pushed for steeper and faster emissions reductions, the United States, China, India, Brazil, and South Africa assumed the role of laggards on this score. In contrast to the Kyoto Protocol, COP17 arrived at a commitment to work on a legally binding agreement applicable to all parties. Until such time, the Kyoto Protocol will not end in 2012 but will continue until either 2017 or 2020.

The European Emissions Trading Scheme

The EU, along with various countries, including the United Kingdom and Denmark, developed emissions trading schemes. European countries, acting in concert through the European Union, formed the Council of the First Action Programme on the Environment, which became involved in the mid-1980s in the formation of the Vienna Convention on the Protection of the Ozone Layer and the Montreal Protocol. The EU has campaigned since 1989 for CO_2 stabilization by establishing reduction targets and timetables. The European Commission proposed a harmonized energy/carbon tax for the European Union during the 1990s. The measure failed due to opposition from Environment Commissioner Carlo Ripa di Meana and certain governments, including the United Kingdom, as well as corporate interests that argued that a tax would undermine their competitive position (O'Riordan and Jordan 1999:84). The corporate lobby groups were represented primarily by the Union of Industrial Employers' Confederations in Europe (UNICE), which opposed a carbon tax at both member state and EU levels (Ellerman, Convery, and de Perthuis 201016). The European Commission reportedly "first discussed the emissions trading scheme as part of its post-Kyoto strategy in 1998" (Gilbertson and Reyes 2009:28).

Oliver Tickell (2008:49–50) delineates the following phases in the EU Emissions Trading Scheme:

- Phase 1 (2005–2007), which applied only to emissions from the largest greenhouse gas emitters, such as power stations and other major industrial facilities responsible for about 40 percent of all EU emissions
- Phase 2 (2008–2012), during which period more EU allowances or permits would be auctioned rather than grandfathered, a policy that has been extended to cover medium-sized emitters
- Phase 3 (post-2012), in which the scheme will be extended, regardless of the outcomes of further COP meetings
- Phase 4 (post-2020)

The EU ETS was implemented in January 2005 and purportedly functions as Europe's most important climate change mitigation strategy. It permits member states to grant their largest industries the right to emit a specific amount of CO_2. The notion of emissions trading began in Germany during the 1970s. The German chemical industry opposed compulsory emissions trading on the grounds that voluntary agreements had operated well at the national level in the past (Brunner 2008:502). In December 2000, the German government created a working group on tradable permits. In contrast to the United Kingdom, Germany never developed a national ETS. Each EU member state sets a cap on overall CO_2 emissions and allocates emission permits to individual industrial facilities, which in turn can either reduce their emissions or purchase permits from facilities with an excess of permits. In 2005, industrial facilities obtained 80 million tons, or 4 percent, more permits than they actually required, resulting in a destabilized emissions market. According to Schreuder (2009:133–34), "the electricity sectors have been accused of taking windfall profits during the first phase of the EU ETS (2005–2007), as they can pass on the cost of carbon permits even when carbon-neutral electricity is produced." While member states had the option to auction up to 5 percent of their allowances, most opted not to do so (Ellerman, Convery, and de Perthuis 2010:50).

The EU ETS covers the following types of operations: (1) combustion operations with a rated thermal input exceeding 20 MW; (2) petroleum refineries; (3) coke ovens; (4) iron production and processing; (5) mining; (6) glass manufacturing installations; (7) ceramic productions installations; and (8) pulp and paper production plants (Brohé, Eyre, and Howarth 2009:113). Reportedly, the "740 biggest emitters (7 percent) covered by the scheme account for 80 percent of the emissions, while the 7400 smallest emitters account for less than 5 per cent of the emissions" (Brohé, Eyre, and Howarth 2009:114). Beginning in 2013 the EU ETS will include CO_2 emissions from petrochemical, ammonia, and aluminum production facilities along with nitrous oxide emissions produced by the aluminum industry (Brohé, Eyre, and Howarth 2009:131). Unfortunately the EU ETS does not cover about 60 percent of all EU CO_2 emissions, including ones from transport, dwelling units, agriculture, low-energy-intensive industries, commerce, and the public sector, including military operations (Brohé, Eyre, and Howarth 2009:142). From 2012 on emissions from all commercial airlines flying in and out of EU airports will be regulated by the Aviation Trading Scheme, which will be part of the EU ETS (Ellerman, Convery, and de Perthuis 2010:264–65).

CO_2 emissions in 2007 were 8.3 percent greater than the 2005 verified 2298 millions tons of CO_2 emissions produced by EU countries in 2005 (Friends of the Earth Europe 2010b:3). The average CO_2 emissions is 2 percent lower for the 2008–2012 phase in large part due to the Global Financial Crisis. But "in seventeen out of twenty-member states—including France, Poland and the UK, 2012 are still higher than measured emissions in 2005" (Friends of the Earth Europe 2010b:133).

The price of carbon under EU ETS fluctuated widely in the initial years of the scheme, varying from a high of 30 Euros for a ton of carbon to a low of 0.03 in December 2007. In April 2006 the carbon price crashed due to the overallocation of emissions permits. In commenting on the price of 15 Euros for a ton of carbon in October 2010, Jacobs (2010:12–13) maintains that

> Such a price is simply not providing enough incentive for energy and industrial companies to invest in low-carbon technologies such as radical energy efficiency, wind and biomass. Indeed, it does precisely the reverse, encouraging the conventional high-carbon investments of gas and coal-fired generation. With Europe embarking on a wave of investment in energy supply that will last up to forty years, this threatens to lock in high European emissions for decades. In turn, this will make Europe's long-term commitment to cut its emission by 80 per cent by 2050 almost impossible to achieve.

Various polluting companies have earned huge windfall profits at times as a result of fluctuating permit prices and have collected permits cheaply when prices have been low (Chivers 2009:137). Some companies were granted so many permits that they were able to emit more CO_2 than previously. Arcelor Mittal, a steel manufacturer, earned 108 million Euros during 2007 to 2009, and Lafarge, a cement manufacturer, earned 142 million Euros in 2009 (Coelho 2011:19). Companies can also purchase carbon credits from the CDM.

In January 2008, the European Union announced a scheme that purportedly would result in 20 percent emissions reductions by 2020 compared with the Kyoto Protocol 1990 baseline. Some EU members failed to achieve their emissions reductions goal for the first round of the Kyoto Protocol and have been instructed by the European Union to make a more concerted effort to do so. Furthermore, the recent Global Financial Crisis has prompted the heads and environment ministers of EU member states to demand that their caps be lowered (Frank 2009:34). Yda Schreuder (2009:160) argues that the EU ETS has fallen short of its alleged aims in that it failed to keep CO_2 emissions reductions on target during the first phase (2005–2007); it remains uncertain whether overall commitment of 8 percent reduction during the second phase (2008–2012) will be met; and carbon leakage, a process by which corporations shift their operations from EU countries to non–Annex I countries or the United States and thereby actually increase greenhouse gas emissions, potentially constitutes as serious problem. According to Andrew Dessler and Edward Parson (2010:163), in contrast to phase one (2005–2007) and phase two (2008–2012), "plans to move toward auctioning in the third phase (2013–2020), from 2013 to 2017, met forceful opposition, and compromise that reduced initial auctioning to 20 percent was necessary to secure adoption of the new Climate Plan." Furthermore, the European Union is negotiating to include aviation and shipping emissions in its ETS and to "implement a border carbon tax to limit competitive losses if the EU's trading partners fail to take similarly stringent action" (Dessler and Parson 2010:163).

The EU has been promoting biofuels as a climate change mitigation strategy. Wall (2010:37–38) reports,

> One EU Directive states that 5.75 per cent of vehicle fuel in the EU must come from biofuels. The biggest single source of biofuels for the EU and a major source for the US is Colombia. However, much of the land used to grow biofuels in Colombia has been taken from local people by right-wing paramilitaries.

The EU plans to link its ETS with existing or planned schemes in New Zealand, Japan, Australia, and the United States.

Other Regional Climate Regimes

While the European Union constitutes the most notable regional climate regime, some other regional governmental bodies function at least as climate regimes. A noteworthy example is the Asia-Pacific Partnership on Clean Development and Climate (often simply referred to as AP6), which consists of the United States, Australia, China, Japan, India, and South

Korea, countries that account for about half of the world's CO_2 emissions (Lowe 2005:178). The Bush administration took the lead on the creation of AP6 and the Howard government in Australia quickly jumped on board in an effort to deflect criticisms of their respective countries' failure to ratify the Kyoto Protocol. Although China, India, and South Korea had ratified it, promises of funding for clean technology investments convinced them to join AP6. Despite some reluctance, Japan joined the pact, probably in order not to be left out of a bloc consisting of powerful Asian economic powers. The partnership eschews setting emissions targets but rather "aims to produce forms of cooperation that facilitate investment in clean technologies, goods and services, accelerate the sharing of energy-efficient best practices, and identify policy barriers to the diffusion of clean energy technologies" (Newell and Paterson 2010:30). In referring to the theme of AP6's first meeting in Australia in January 2006, then Australian prime minister John Howard said, "Our societies require of us that we find solutions to these issues [climate change and increased greenhouse gas emissions] that maintain the momentum of economic growth" (quoted in Ponting 2007:407).

The Group of 77, which now consists of over 130 developing countries, tends to be dominated by its larger members, such as China and India, which contend that emissions reductions would hinder their economic development and efforts to eradicate poverty. The Organization of Petroleum Exporting Countries tends to act informally in coordinating its members' positions in climate change negotiations, but as a group they are committed to the "protection of their main economic export, oil, and prevention of any treaty that undermines the significant usage of fossil fuels" (Maslin 2004:131). The Alliance of Small Island States was established in 1990 and is a coalition of some 43 low-lying small island states, some of which also belong to the Group of 77, that are highly vulnerable to sea level rise resulting from climate change. Included in this body are the Maldives, Mauritius, Singapore, Papua New Guinea, the Federated States of Micronesia, Tuvalu, the Marshall Islands, Vanuatu, Tonga, Samoa, Nauru, Fiji, and the

Cook Islands. As Mark Maslin (2004:121) observes, "The AOSIS position has always been to get the tightest control on global emissions as their countries seem to be the most at threat from the impacts of global warming." Conversely, tensions exist within AOSIS, such as "between Singapore, which is wealthier in per capita terms than some OECD countries, and which is seeking to avoid an emissions reductions target under a successor agreement to Kyoto Protocol, and other SIDS [small island development states] who argue that wealthy countries such as Singapore should reduce their emissions" (Barnett and Campbell 2010:102). Furthermore, some larger Pacific island states, in particular Papua New Guinea, are seeking recognition for emissions reductions by regulating deforestation. Some observers are skeptical about whether this will contribute to a significant reduction of emissions.

National and State or Provincial Climate Regimes

Various countries, as well as states or provinces of countries, have created emissions trading schemes. Countries include Denmark, the United Kingdom, Norway, Switzerland, Canada, Japan, New Zealand, and South Korea (Newell and Paterson 2010:106; International Energy Agency 2010:12). States and provinces include New Jersey, Massachusetts, New Hampshire, Oregon, Florida, California, New Mexico, and New South Wales (Australia). Regional ETSs include the New England–Canada consortium, the Western Climate Initiative, which includes 11 US states and Canadian provinces and is to take effect in 2012.

The United States

Senators Al Gore (D-TN) and Tim Wirth (D-CO) introduced the issue of climate change in congressional hearings in 1987 and 1988 (Bryner 2001:141). James Hansen, the director of NASA's Goddard Institute for Space Studies, testified before Congress during the hot summer of 1988 about the need to act upon the dangers of anthropogenic climate change; Congress essentially ignored his advice. The George H. W. Bush administration opposed binding limits on greenhouse gas emissions during the UN Conference on the Environment and Development in 1992 and successfully pushed for voluntary reductions in the creation of the FCCC (Bryner 2001:142). In 1993 the Clinton administration created a Climate Change Action Plan following FCCC dictates that proposed a reduction of greenhouse gas emissions to 1990 levels by 2000, an objective that never was fulfilled. The United States negotiated at Kyoto credits for tree planting since trees sequester CO_2. It opposed carbon taxation but successfully campaigned for the inclusion of an ETS in the Kyoto Protocol. The Clinton administration "hoped that its trading *über alles* position would help it to win over Republican Senators who liked markets, but detested regulation" (Driesen 2010:134). The Clinton administration signed the Kyoto Protocol and agreed to reduce greenhouse gas emissions by 7 percent, but ratification of the protocol was defeated in the US Senate by a vote of 95–0.

While running for president in 2000, George W. Bush claimed that he supported an antipollution scheme that would include mandatory caps on CO_2 emissions from utility companies (Brohé, Eyre, and Howarth 2009:157). However, Bush shifted his stance to support for voluntary schemes. In February 2002, his administration launched a Global Climate Initiative that proposed to "reduce the GHG intensity of the US economy, measured as GHG emissions per unit of total gross domestic production" (Brohé, Eyre, and Howarth 2009:158). Various members of the Bush administration had strong ties with the fossil fuels industry, including Vice President Dick Cheney, Condoleezza Rice, and Donald L. Evans. According to Brian Black and Gary Weisel (2010:106),

> The Bush administration sharply reduced NASA's earth science budget and appointed its own people to key NASA positions. The appointees often had limited scientific expertise but maintained ideological positions that were acceptable to the administration. There are a number of documented cases in

> which senior NASA managers (such as George Deutsch) and officials in the White House (such as Philip Cooney) rewrote scientific reports to the government such that they would highlight scientific uncertainties and minimize the danger presented by global warming.

The Bush administration placed pressure on the Environmental Protection Agency in summer 2003 to alter its 600-page report by editing out all references to potential dangerous impacts of climate change on the United States. It unsuccessfully ordered James Hansen not to speak in public forums about the grave implications of climate change. During the George W. Bush era, various cap-and-trade bills had been submitted in the US Congress, including the McCain-Lieberman Climate Stewardship and Innovation Act in the US Senate in 2007 and the Kerry-Snowe Global Warming Reduction Act in the US Senate, also in 2007, both of which were defeated.

Some US-based multinational corporations have expressed concern about how to contain the risks and maximize profits from climate change and formed the US Climate Action Partnership, which in 2007 called upon the US government to "quickly enact strong national legislation to require significant reductions of greenhouse gas emissions" (quoted in Smith 2011:23). In 2008 membership of the group included the American International Group, Boston Scientific Corporation, Chrysler, ConocoPhillips, Deere & Company, Dow Chemical, Ford, General Motors, Johnson & Johnson, PepsiCo, Rio Tinto, Shell, Siemens, and Xerox, along with the National Wildlife Federation and the Nature Conversancy. In the wake of COP15 in December 2009 and the Climategate scandal, ConocoPhillips, BP America, Caterpillar, and Xerox left the US Climate Action Partnership.

On January 27, 2009, President Barack Obama stated,

> I will reverse our dependence on foreign oil while building a new energy economy that will create millions of jobs. . . . America's dependence on oil is one of the most serious threats that our nation has faced. It bankrolls dictators, pays for nu-

> clear proliferation and funds both sides of our struggle against terrorism. (Quoted in Caffentzis 2010:564)

His administration has been promoting an emissions trading scheme, carbon capture and storage technology, stricter car gas-mileage standards, biofuels, and nuclear energy as alleged climate change mitigation strategies. Furthermore, much of the Obama administration's planned spending is directed to new roads and fossil fuel power plants. Todd Stern, Obama's chief climate negotiator, claims that it is impossible for the United States to aim for 25 to 40 percent cuts in greenhouse gas emissions by 2020, despite the fact that the IPCC maintains that developed countries need to aim for 25 to 40 percent cuts by then to avoid dangerous climate change. Steven Chu, the secretary of energy in the Obama administration, appears to be a big fan of geo-engineering, including "painting the roofs of buildings and road surfaces white to reflect sunlight back into space" and "burning nuclear waste in special reactors that will transmute it into more benign elements" (Perlmutter and Rothstein 2011:166).

The US House of Representatives passed the American Clean Energy and Security Act in June 2009, which calls for a mandatory national emissions reduction of 17 percent below 2005 levels (or the equivalent of about 4 percent below 1990 levels) by 2020 (Christoff 2010:650). The bill includes a cap-and-trade scheme that reportedly is to cover 85 percent of US greenhouse gas emissions, including those from the power, industry, transport, commercial, and residential sectors. The targets are set against 2005 emission levels, at 3 percent by 2012, 17 percent by 2020, 42 percent by 2020, and 83 percent by 2050. A similar bill, however, failed to obtain the requisite 60 percent vote for passage in the Senate (International Energy Agency 2010:14). In response to the Senate bill, the American Petroleum Institute convened a series of public rallies opposing it, in large part because the bill allocated 35.5 percent of the free permits to the power or utility sector and only 2.25 percent to petroleum refiners (Perlmutter and Rothstein 2011:158).

China

China has been extremely reluctant to set firm greenhouse gas emissions targets and timetables. While its leadership claims to be committed to curtailing greenhouse gas emissions, they assert that the country's primary objective is economic development and the eradication of poverty within its borders. Between 1978 and 2004, China improved its energy efficiency from 26 percent to 33 percent (Longhai 2008:216). In terms of steel production, energy consumption per ton had been reduced by over 30 percent. In 1990 China established the National Coordination Panel on Climate Change as a unit of the State Environmental Protection Commission (Zhang 2003:67). While not a signatory of the Kyoto Protocol, China is an active participant in the Clean Development Mechanism. In 2005 China passed the Renewable Energy Law, which made a commitment to the development of wind, solar, geothermal, and biomass (Longhai 2008:216).

Pan Yue published the first official report on China's "Green GDP" in 2006 (Leonard 2008:42). The report indicated that air, water, and solid-waste pollution resulted in US$64 billion in damage in 2004, accounting for 3.05 percent of GDP that year. In June 2007 China published its first national climate change strategy, *China's National Climate Change Programme* (National Development and Reform Commission 2007). The report stressed China's commitment to economic development. While admitting that its average 9.5 percent annual growth has contributed to a quadrupling of greenhouse gas emissions since 1979, the report noted that China's cumulative historical emissions and per capita emissions rate remain low compared with those of developed countries. While rejecting any international agreement that would attempt to curb China's economic growth, the report delineated three components to the country's climate change mitigation strategy: (1) reducing its energy intensity by 20 percent during 2006 to 2010; (2) renewing its commitment to reforestation and forest management program as forms of CO_2 sequestration; and (3) viewing its "one-child-only" policy as a climate change mitigation strategy. A 2009 report describes massive plans to reduce China's energy intensity (National Development and Reform Commission 2009).

Despite its establishment of a voluntary national climate regime, China's greenhouse gas emissions are rising at more than 5 percent per annum, in large part due to its "burgeoning economy, rapid expansion of coal-generated energy and a cement-fueled construction boom" (Hulme 2009:273). Shortly before the 2009 UN FCCC Copenhagen conference, the Chinese government set its first carbon target, an intensity one that would slow the rate of greenhouse gas emissions rather than cut them. Under this plan, China expects to remain the world's biggest emitter of greenhouse gas emissions.

On the positive side, China has made significant headway in the development of wind and solar power. Between 2005 and 2009, its wind power generation capacity doubled annually (Watts 2010:276). Furthermore, China has taken the lead in making and exporting the most photovoltaic panels and has "launched a programme to install millions of solar heaters and mulled feed-in tariff incentives to further promote solar power" (Watts 2010:276). It also is in the process of developing a "zero-carbon city" in Dongtan near Shanghai (Leonard 2008:43).

Conclusion

Climate change constitutes an internal problem that will require agreement by the majority of the countries in the world, particularly those that are the largest emitters of greenhouse gases, such as the United States, Canada, the EU countries, Russia, Japan, China, India, Brazil, and even Australia, with a population of only some 22 million people. All climate regimes ranging from the Kyoto Protocol to the European Union's emissions trading scheme to national regimes face serious implementation and accountability problems. Achim Brunnengräber (2006:226) assesses the limitation of the climate regime approach manifested in the Kyoto Protocol as follows: "It has not yet been possible to specify and implement the mechanisms for the reduction of greenhouse gases in such a way that a reduction of CO_2 measured in absolute figures can be guaranteed." Even if the targets of the Kyoto Protocol are fully met, they will only result in modest greenhouse gas emissions reductions. Climate change conferences and climate regimes repeatedly recommend or establish targets for reducing greenhouse gas emissions, particularly CO_2. For example, the World Conference on Climate and Development held in Hamburg in 1988 called for developed countries to "commit themselves to reducing their [carbon dioxide] emissions by at least 30 per cent by the year 2000 and 60 per cent by the year 2015, based on 1986" (quoted in Falk and Brownlow 1985:195). Developed countries did just the opposite: they generally increased their CO_2 emissions.

Furthermore, Ray Kiely (2007:129) argues that conventional climate regimes "are too easily guilty of ignoring the uneven development of international capitalism, and therefore the unequal context in which rights, values, ethics and international institutions operate." It has become increasingly obvious, as Gert Spaargaren and Arthur Mol (2008:351) observe, that national environmental regimes "fall short due to the growing organisational and technical complexity of globalising production and consumption systems." In terms of the political economy of research funding, James Hansen (2007:31) has gone so far as to suggest that those climate scientists who tend to downplay the dangers of climate change are more likely to obtain funding.

5

Why Green Capitalism Is Insufficient to Mitigate Climate Change

Corporations by and large initially tended to ignore, deny, or downplay the reality of anthropogenic climate change or ascribe it to strictly natural forces. A 2007 Union of Concerned Scientists report titled *Smoke, Mirrors & Hot Air: How ExxonMobil Uses Big Tobacco's Tactics to "Manufacture Uncertainty" on Climate Change* indicates that ExxonMobil contributed nearly $16 million between 1998 and 2005 to a network of 43 contrarian organizations. The Global Climate Information Project was a coalition of business, labor, and farm groups in the United States, the activities of which are described in greater detail toward the end of this chapter.

Corporations, including ones part and parcel of the fossil fuels industry, have increasingly shifted from a stance of climate skepticism, as embraced by the now defunct Global Climate Coalition (GCC), to accepting the reality of anthropogenic climate change and asserting that as responsible corporate citizens, they can contribute to *sustainable development* as well as climate change mitigation. The World Council for Sustainable Development consists of over 100 corporate members, including Renault, Fiat, Shell, Texaco, BP, Mitsubishi, and Toyota. It claims to be committed simultaneously to solid financial performance and environmental sustainability. In addition to making a pitch for sustainable development, green capitalism emphasizes schemes such as emissions trading, carbon offsets, and energy efficiency. It does not address the issue of social justice or equity and ultimately privileges profit making and economic expansion over environmental sustainability. Green capitalism may embrace various technological fixes and alternative energy sources, ranging from nuclear power to renewable energy sources, such as solar energy, wind energy, geothermal energy, wave or tidal energy, and hydropower. While many of these are important in climate change mitigation, the more or less exclusive focus on them comes under the rubric of *ecological modernization*, a stance that many environmentalists (including ones in the climate movement) embrace. Although ecological modernization concedes that environmental problems may be a by-product of global capitalism, it rejects transcending this system and rather advocates reforming it by introducing various technological innovations and energy efficiency. From this perspective, capitalism can be made more "environmentally friendly" through environmental regulations and technological changes managed by ecologically sensitive governments working in concert with corporate interests.

Green Capitalism and Environmental Sustainability

Green capitalism has come to embrace the notion of *sustainable development* that was introduced in the 1987 report of the World Commission on Environment and Development, often referred to as the Brundtland Report in recognition of its chairperson, former Norwegian prime minister Gro Harlem Brundtland. The report recognized that the global economy is depleting natural resources at an astounding rate, which cannot be maintained indefinitely. It went on to argue that economic growth is necessary to deliver prosperity to the developing world, but it has to be sustainable. Bearing these thoughts in mind, the commission defined sustainable development as "development that meets the needs of the present without compromising the ability of future generations to meet their own needs" (World Commission on Environment and Development 1987:326). The commission, however, also asserted that "technology and social organization can be both managed and improved to make the way for a new era of economic growth" (World Commission on Environment and Development 1987:8). Thus, sustainable development implies the possibility of combining economic expansion and environmental sustainability.

At the 1992 Earth Summit, the Business Council for Sustainable Development released *Changing Course* delineating a vision of corporate environmentalism based on four premises (Karliner 2000:179):

- Ongoing economic growth based upon market forces is a basic prerequisite for sustainable development.
- Pricing mechanisms could correct the environmental distortions emanating from the global economy.
- Voluntary regulation constitutes the most effective means for making businesses more environmentally sustainable.
- Technological and managerial innovations would result in cleaner productive processes and more efficient utilization of resources.

The concept of sustainable development remains a contested one and means different things to different parties. Julian Agyeman (2005) defines sustainable development in terms of the "three E's":

- Economy: creating economic activity that results in decent employment, income, and a tax base
- Ecology: protecting a city's or community's natural assets and creating a less polluted environment
- Equity: ensuring that all people have access to economic opportunities and are not environmentally endangered due to their social class

According to Agyeman (2005:43), "Sustainability is at least as much about politics, injustice, and inequality as it is about science or the environment." While I more or less agree with this assertion, he appears still to accept the notion of ongoing economic growth or expansion. However, as anthropologist Alf Hornborg (2001:9) so aptly argues, "as long as the concept of development continues to hinge on growth, the notion of 'sustainable development' remains an oxymoron."

Climate Capitalism

Peter Newell and Matthew Paterson (2010:9) call for what they term *climate capitalism* but recognize that it will not be easy to achieve:

> So the challenge of climate change means, in effect, either abandoning capitalism, or seeking to find a way for it to grow while gradually replacing coal, oil and gas. Assuming the former is unlikely in the short term, the questions to be asked are, what can growth be based on? What are the energy sources to power a decarbonised economy? . . . What kind of climate capitalism do we want? Can it be made to serve desirable social, as well as environmental, ends?

In essence, they seek to decouple increasing greenhouse gas emissions from economic expansion, an objective that strikes me as highly dubious and naively wishful in an effort to save capitalism from itself.

In May 1997, Sir John Browne, the CEO of British Petroleum, admitted that fossil fuels were contributing to greenhouse gas emissions (Camilleri and Falk 2010:305). Shortly afterward, various other energy companies, including Texaco, Royal Dutch Shell, and Sun Oil, made similar admissions. According to Camilleri and Falk (2010:304),

> By 1998 the energy giants and other leading corporations had decided to attend COP-4 meetings and to make formal presentations on their planning for the transition from fossil fuels. In 2001, a Fortune poll of 5000 US business executives found that 75 per cent of the respondents considered global warming to be a serious problem.

The US Climate Action Partnership includes BP America, Shell, ConocoPhillips, Ford Motor Company, General Electric, General Motors, Chrysler, Deere, Caterpillar, Dow Chemical,

DuPont, Johnson & Johnson, Pacific Gas and Electric Company, Alcoa, and Siemens. In 2007 it called for a government-imposed mandatory carbon emissions target for its own industries (Craven 2009:115). The CEO of Royal Dutch Shell recommended in 2006 a movement to a "low-carbon economy through cap-and-trade policies" (Craven 2009:119). The International Chamber of Commerce advocates the adoption of technologies that serve to mitigate climate change (Gonzalez 2005:356). Various corporations, including Boeing, Lockheed-Martin, Toyota, Whirlpool, 3M, BP, Sun Oil, and American Electric Power, have lent their support to the Pew Center on Global Climate Change (established 1998), which views technological innovation as the first step in addressing climate change (Hamilton 2007:85). In 2011 the Pew Center became the Center for Climate and Energy Solutions (www.c2es.org). The US Climate Action Partnership claims to be "committed to a pathway that will slow, stop and reverse the growth of US emissions while expanding the US economy" (quoted in Newell and Paterson 2010:36). Its membership includes major greenhouse gas emitters such as Alcoa, BP, Ford, General Electric, and Shell.

The International Chamber of Commerce advocates the implementation of technologies that serve to mitigate climate change. Corporations such as Shell, BP, DuPont, Morgan Stanley, and others established the Partnership for Climate Action and the Business Environmental Leadership Council. Gareth Dale (2007:118) delineates seven categories of climate change mitigation strategies that many multinational corporations are embracing: (1) investing in renewable energy sources, (2) improving energy efficiency, (3) encouraging employees and consumers to reduce their personal emissions, (4) developing technologies for sequestering CO_2, (5) shifting to biofuels, (6) carbon offsetting, and (7) influencing national and international climate policy making. In reality, corporations want it both ways—namely, metaphorically not to kill the goose that laid the golden egg but to continue on with their commitment to profit making and economic expansion. In a similar vein, Dale (2007:131) argues,

> The climate change strategies of corporations and states involve an attempt to square two conflicting imperatives. To avoid instability and maintain the conditions necessary to capitalism's reproduction, climate chaos must be mitigated, but, for state and corporations alike, this must not affect the button line of maximising profitability and outcompeting rivals.

Victor Wallis (2010:22) asserts that the notion of *green capitalism* is an oxymoron in the sense that is seeks to combine two contradictory notions—namely, to "prioritize the health of the ecosphere, with all that this entails in terms of curbing greenhouse gases and preserving biodiversity," while at the same time promoting "growth and accumulation, treating both the workforce and the natural environment as mere inputs." In reality, some corporations, such as insurance and renewable energy companies, and corporate interest groups, such as the European Wind Energy Association, stand to benefit financially from acknowledging the reality of climate change. Martyn Turner and Brian O'Connell (2001:163) boldly assert that the battle against global warming "will be fought in corporate finance departments of investment banks, in offices of venture capitalists, and in R&D departments of major corporations."

Conventional Economists' Take on Climate Change

Nicholas Stern (2007) has proposed the most comprehensive of the neoliberal approaches for mitigating climate change. In the first page of his report, he states, "Climate change presents a unique challenge for economics: it is the greatest example of market failure we have ever seen." Stern (2007:xv) argues "that if we don't act, overall costs and risks of climate change will be equivalent to losing at least 5% of global GDP each year, now and forever." He asserts that the "costs of action—reducing greenhouse gas emissions to avoid the worst impacts of climate change—can be limited to around 1% of global GDP each year," a course of action that would lead to stabilization at around 500 to 550 ppm CO_2 by 2050 (Stern 2007:xv). He maintains that all countries need to be involved in a massive effort to mitigate climate change while none of them, whether developed or developing, need to "cap aspirations for growth" (Stern 2006:xvii). Stern proposes three main strategies for reducing greenhouse gas emissions: (1) reducing the demand for carbon-intensive products, (2) improving energy efficiency, and (3) shifting to lower-carbon technologies.

More recently, Stern (2009:54) has argued that spending "2% of GDP per annum is well worth [it] to reduce the chances of temperature increases above 5°C from around 50% to about 3%." He claims that stabilizing GHEs at 400 ppm CO_2e is unrealistic because humanity has already reached a level of 430 ppm CO_2e (around 380 ppm CO_2) and will be at 450 ppm CO_2e within a decade (CO_2e is a measurement that includes all greenhouse gas emissions) (Stern 2009:150). David King, who was the United Kingdom's chief scientific advisor when the Stern Report was released, asserted that 550 ppm of CO_2e was a more realistic target (Simms 2007:207).

Stern and other conventional economists, such as Ross Garnaut (2008) in Australia, who engage in cost-benefit analyses of climate change mitigation, tend to refer to "social discount rates" that seek to factor in the reality that the "benefits (i.e., avoided damages) of today's investments in mitigation activities (i.e., that lead to reductions in greenhouse gas emissions) will only be realised by future generations" (Hulme 2009:121). Higher discount rates generally are set at 4 percent per year or more, whereas lower ones are set at 3 percent or lower.

Mike Hulme (2009:121) asserts that with regard to social discount rates,

> any damages from climate change will increasingly be borne by future generations. If those people were to speak to us today, they would ask us to forgo some of our own consumption so that they could enjoy the future benefits of a reduced change in climate. In other words they would ask us to divert some investment today into climate change mitigation, thereby reducing, by some amount, future damage from climate change.

William Nordhaus (2008) advocates a go-slow approach to curbing greenhouse gas emissions in that he acknowledges that climate change is largely anthropogenic and that it is necessary to make a slow transition away from carbon-emitting energy sources. Nordhaus assigns value to the natural environment and human well-being using conventional economic measures (York, Clark, and Foster 2009:6). He argues for a "climate-policy ramp" in which modest reductions in greenhouse gas emissions in the near term would be followed by more ambitions reductions in the long term.

Nick Dallas (2009) in *Green Business Basics* delineates six categories of solutions for reducing greenhouse gas emissions. The first of these entails reducing consumption of goods and services, which he maintains is "very contentious as it implies reduced economic activity and is difficult to sell as the global economic system is based on increasing consumption and annually increasing economic growth." (Dallas 2009:53). The second solution advocates modifying consumption, which Dallas (2009:53) sees as being "more palatable for most and does not necessarily lead to reduced economic activity, only less energy-intensive activity, such as shifting to vegetarianism and sourcing locally-produced food." Indeed, over the past several years, many books have appeared that advocate green consumerism and instruct individuals as to how they can help prevent climate change (Langholz and Turner 2003; Goodall 2007; Steinman 2007). Dallas (2009:57–59) also writes approvingly of solutions such as improving energy efficiency and optimizing the energy production mix, developing low-carbon and renewable sources of energy, sequestrating carbon dioxide, and educating people about the implications of their behavior in terms of generating greenhouse gas emissions. He doubts that biofuels in most instances are a likely substitute for oil and views nuclear energy as problematic (Dallas 2009:62–65). Dallas (2009:93) does admit that governments have an important role to play in climate change mitigation.

Servaas Storm (2009:1022–26) delineates a typology of approaches to climate stabilization, which he terms *quadrants*:

1. Quadrant 1: Climate change can be halted by using market mechanisms such as carbon markets.
2. Quadrant 2: This includes green Keynesian, which favors a global carbon tax combined with direct regulation.

3. Quadrant 3: This includes a high global tax and public investment in energy conservation, photovoltaic installations, organic agriculture, and public transport (Lohmann 2009; Speth 2008). This would entail low, no, or even degrowth (at least in developed countries) and a drastic reformation of corporations, income distribution, and so forth.
4. Quadrant 4: This is an anticapitalist stance rooted in the "climate justice movement, Green socialist and feminist politics, and (deep) ecological and anti-globalization movements" that advocates collective ownership of natural resources, democratic management of production, and low consumption.

While a growing number of mainstream economists have come to realize that climate change could potentially "kill the goose that laid the golden egg," they almost always are wedded to the belief that global capitalism needs to continually expand economically. As a result, as Foster (2009a:12) observes, "orthodox economists constitute the leading ideological opponents of aggressive reductions in greenhouse gas emissions, even at the risk of a planetary inferno—and that is their primary role as ideological defenders of the capitalist system and promoters of its drive for profits and accumulation at any cost." Foster, Clark and York (2009:1088) contend that orthodox economists' treatment of the environment constitutes a "Midas effect" that "constantly seeks to transmute ecological values into economic ones." Mainstream economists assume that the market will ultimately safeguard environmental values by placing a price on the planetary ecosystem and its components, including water, air, fauna, flora, and presumably human beings themselves.

The G8 and the World Bank on Climate Change

The Group of 8 (G8) constitutes a political forum and annual meeting attended by the heads of state of eight industrial powerhouses: the United States, Canada, the United Kingdom, Germany, Japan, Italy, France, and Russia. The Group of 7 preceded the present G8, which added Russia in 1997. The G8 countries comprise about 13 percent of the world's population but are responsible for about 25 percent of the world's greenhouse gas emissions. While the G8 does not have any official decision-making powers, its members have considerable influence over various global governance bodies, including the UN Security Council, the World Trade Organization, the International Monetary Fund, the World Bank, and the Organization for Economic Cooperation and Development. In 2000 the G8 created a Renewable Energy Task Force with the aim of achieving a technological solution to climate change (Carbon Trade Watch 2005:6). While serving as chairperson of the G8 meeting in 2005 in Scotland, UK prime minister Tony Blair called for the G8 to adopt a strong stance on climate change mitigation.

The G8 has recognized that human activities "are now causing atmospheric concentrations of greenhouse gases—including carbon dioxide, methane, tropospheric ozone, and nitrous oxide—to rise well above pre-industrial levels" (quoted in Jarman 2007:5). At the 2005 G8 meeting, George W. Bush finally abandoned his assertion that climate science is incorrect in ascribing most climate change to anthropogenic activities. The World Economic Forum issued a statement on June 2006 supporting Blair's call at the G8 to address climate change. Despite the G8's purported commitment to climate change mitigation, its 2007 declaration in Rostock, Germany, indicates a continuing emphasis on economic expansion as indicated by its acceptance of ongoing motor vehicle production (Bello 2007). The G8 proposed promoting the development of nonfossil fuels, such as hydrogen, for new cars. At the G8 summit in Japan in July 2008, the eight state leaders agreed to reduce CO_2 by 50 percent by 2050 but failed to specify a starting year (Li 2009:1045). The state leaders of Brazil, China, India, Mexico, and South Africa rejected this proposition and urged the developed countries to reduce emissions by 80 to 95 percent from 1990 levels by 2050 and requested financial support to assist developing countries in adapting to climate change. Hu Jintao, the president of China, insisted that his country as a developing country needs to concentrate on industrialization and improving people's living standards. India indicated that the only acceptable limit to greenhouse gas emissions that it would accept was the same per capita amount presently being emitted by developed countries (Perlmutter and Rothstein 2011:163).

In 2005 the G8 asked the World Bank to develop a climate change mitigation framework. Ironically, the World Bank has been promoting the utilization of fossil fuels (Jarman 2007:68). In 2000 the Bank initiated an Extractive Industries Review to determine whether its investments in petroleum, natural gas, and mining are contributing to climate change. The review concluded in 2004 that many of the Bank's energy projects have contributed to environmental degradations, social upheaval, and conflict. However, the Bank concluded that continuing with its extractive energy projects will also ultimately play a significant role in reducing greenhouse gas emissions created by the fossil fuels industry. It committed to increasing its investments in renewable energy sources by 20 percent per year. Friends of the Earth conducted a study of World Bank Group agencies that showed their funding for renewable-energy and energy-efficiency projects increased only 7 percent in 2005 (Jarman 2007:71). The World Bank also created the Prototype Carbon Fund in 1999, which permits the purchase of emissions reductions credits under the Clean Development Mechanism (CDM) and Joint Implementation (JI) (Newell and Paterson 2010:79). JI entails trading credits between Annex I and non–Annex I parties in order to meet their commitments of emissions reduction targets and allows an Annex I party to host a project that generates emissions reduction credits that can later be purchased by another Annex I party.

Climate Mitigation Strategies of Oil Companies

The international oil industry is divided between private and state oil companies, the latter of which reportedly control about 70 percent of oil and natural gas reserves (Lovell 2010:73). Nationalized oil companies exist in Iran, Iraq, Kuwait, Saudi Arabia, the United Arab Emirates, Angola, Brazil, Nigeria, Venezuela, Mexico, and Norway. Nevertheless, private oil companies, particularly those belonging to what had been the "Seven Sisters" (Royal Dutch Shell, Exxon, BP, Gulf, Texaco, Mobil, and

Chevron), now the "Six Sisters" given the merger of Exxon and Mobil, dominate the international oil industry (Lovell 2010:71). Shell, BP, and ExxonMobil alone reportedly account collectively for about 8 percent of greenhouse gas emissions (Newell and Paterson 2010:51). Furthermore, 6 out of the world's 10 richest corporations are oil companies.

Just like capitalism, the oil industry is not monolithic. Bryan Lovell (2010:42–66) refers to an "Atlantic divide in Big Oil" within which US-based oil companies have been more apt to fall into the climate skeptic camp, and European-based ones are willing to admit the existence of anthropogenic climate change and the need to mitigate it. A debate that occurred at an international conference titled "Coping with Climate Change" convened by the UK Geological Society's Petroleum Group in 2003 illustrated this divide, with Frank Sprow, the vice president of BP, taking the latter perspective.

Jon Birger Skjærseth and Tora Skodvin (2009:12–13) delineate three models that seek to explain corporate climate strategies:

1. The Corporate Actor model focuses on "factors such as environmental risk, environmental reputation and organisational learning capacity."
2. The Domestic Politics model is based on the premise that multinational corporations are "heavily influenced by the framework conditions of their home-base countries" and maintains that political institutions can shape corporate practices.
3. The International Regime model emphasizes "how international environmental regimes may trigger changes" in corporate policies and actions.

While ExxonMobil, a US-based company, has claimed to have an overall positive environmental performance record, it opposed the Kyoto Protocol on the grounds that it would prove excessively expensive and unfair to developed nations. Despite the fact that it has given a fair amount of attention to climate change as problematic in its public relations information, it also staunchly supported the now defunct Global Climate

Coalition. Royal Dutch Shell has asserted its commitment to protecting the environment and human populations and fully acknowledged the reality of anthropogenic climate change; it announced in 1998 that it aimed to reduce greenhouse gas emissions from its own operations 10 percent from 1990 levels by 2002 and supported passage of the Kyoto Protocol. Finally, Statoil Corporation, created in 1972 as a Norwegian state oil company but partially privatized in 2001, supported the Kyoto Protocol and adopted a voluntary emissions reduction program. Aside from the degree to which these oil companies may have met their own emissions reductions targets, Skjærseth and Skodvin (2009) argue "that the domestic political context of the company's home-base countries is more significant in explaining differences in corporate climate strategy than are company-specific factors." Given that European countries have been more favorable toward climate mitigation strategies, such as the Kyoto Protocol, than the United States, it is not surprising that both Royal Dutch Shell and Statoil at least gave lip service to their need. Conversely, while Shell dabbled in energy alternatives for a while, in March 2009 the company decided to cease investing in wind, solar, and hydroelectric power on that grounds that these were not economically viable (Newell and Paterson 2010:44–45). Furthermore, despite BP's assertion of being committed to increasing its investments in alternative energy sources, the Global Financial Crisis promoted the company to reduce its alternative energy portfolio.

Mainstream NGOs on Green Capitalism

Many peak environmental NGOs have also come to embrace some variant of green capitalism under which they collaborate with corporations. The prestigious World Resources Institute has more than 30 "corporate supporters," many of which are major polluters, such as Monsanto, Shell, BP, and Cargill Dow (Karliner 2000:190). The National Audubon Society, based in the United States, has accepted funds from Cargill, Chevron, Dow Chemical, DuPont, Ford, Motorola, and Scott Paper. The World Wildlife Fund (WWF) belongs to the European Roundtable of

Industrialists (Lohmann 2006:58). Even Greenpeace "has moved from being critical of corporate lobby groups and carbon trading to complete acceptance" (Lohmann 2006:56). The Climate Action Network has come to embrace the concept of an emissions trading scheme as a potentially effective climate change mitigation strategy.

Li and Minqi (2008:60) discusses the contradictory nature of many people who are involved in the environment movement, which he contends consists, by and large, of upper-middle-class people, such as academics, technicians, managers, financial analysts, and other professionals. He asserts that some of these people could potentially become rapidly radicalized in revolutionary times, such as during the 1960s, but they can also serve as a significant ally of capitalists during counterrevolutionary periods. For instance, many of them acted as supporters of neoliberalism during the 1980s and 1990s and the restoration of capitalism in the former Soviet Union, Eastern Europe, and China. According to Li (2008:61), "The upper-middle-class environmentalists . . . have to put their desperate hope (or faith) in technological miracles on the one hand and the power of moral persuasion on the other hand (which they hope would convince the capitalist class to behave morally and rationally)."

Carbon Trading Markets

Larry Lohmann (2006:89) maintains that "far-sighted companies treat carbon trading as an opportunity to gain *new* property rights, assets and openings for capital accumulation, even if climate change is accelerated in the process." In a similar vein, Newell and Paterson (2010:27) maintain that "emissions trading emerged as the preferred option because of its ideological fit with neoliberal logic. But it was also more successful because of its fit with the interests of newly dominant financial actors."

EcoSecurities in 1997, CO_2e in 2000, and Point Carbon also in 2000 created private emissions markets (Newell and Paterson 2010:27). The Chicago Climate Exchange commenced carbon trading in September 2003. The World Bank is one of the biggest players in the carbon market (Lohmann 2006:341). Carbon

Expo commenced its carbon market trade fairs and conferences in 2004, and there are now various carbon market associations, such as the International Emissions Trading Association, the Emissions Marketing Association, and the Carbon Markets and Investors Association (Newell and Paterson 2010:28).

According to Camilleri and Falk (2010:279), "Businesses joined on a voluntary basis with a view to having a hand in shaping the new carbon market, and positioning themselves in future developments." Various corporations also view their concern about climate change as a manifestation of their social responsibility. Carbon sink credits are popular in the voluntary carbon market. For example, the UK-based Carbon Neutral Company sells carbon credits. Lohmann (2006:172) notes, "Carbon credits go to well-financed, high-polluting operations capable of hiring professional validators of counterfactual scenarios."

Various corporations have turned to green capitalism as a way of making themselves appear to be paragons of corporate social and environmental responsibility. For example, in contrast to its numerous environmental transgressions in the past, Walmart, with assistance from Al Gore, "saw in climate change the opportunity to do some re-branding with a vision that includes powering its facilities and fleet of vehicles with renewable energy, cutting back on waste and selling green products" (Newell and Paterson 2010:53).

Joint Implementation, the Clean Development Mechanism, and Other Offsets

The Kyoto Protocol includes schemes known as the Joint Implementation and the Clean Development Mechanism, both of which allow developed countries to purchase emissions credits from developing countries that have invested in low-emission or energy-reduction projects (table 5.1). According to Melanie Jarman (2007:66), the credits earned in any CDM project are "based on the difference between (1) the amount of greenhouse gases that would have been emitted if the CDM project host had followed a more traditional development project ('the baseline'), and (2) the amount of greenhouse gases generated when

Table 5.1. Clean Development Mechanism Projects in Selected Countries

Country	*Number of Projects*
India	178
Brazil	94
Mexico	78
China	42
Chile	15
Malaysia	14
Honduras	10
South Korea	10
Ecuador	9
Indonesia	8
Philippines	8

Source: See http://cdm.unfccc.int/Statistics/index.html. See Potter et al. (2008:263).

a country follows a low-emission path." Christoph Sutter and Juan Carlos Parreno (2007:75), based on an analysis of officially registered CDMs, conclude that none of them are likely to fulfill the Kyoto Protocol's twofold objective of simultaneously delivering greenhouse gas emissions reduction and contributing to sustainable development. In August 2006, China, Brazil, Korea, and India were hosting over 61 percent of the 265 then existing CDM projects (Lohmann 2006:147).

Wall (2010a:35) offers the following critique of CDM:

> In practice, it often involves paying polluting companies for efficiency investments they would have made without the CDM. Even if this was seen as desirable, there are a huge number of loopholes that can be exploited by firms to milk the system, and there is evidence of thinly disguised fraud. For all these reasons, hypothetical reductions in greenhouse gases rarely occur and the CDM can be used to expand production, which creates more damage to the environment.

Fortunately the Kyoto Protocol does not permit nuclear power to become part of a CDM or JI project.

Reducing Emissions from Deforestation and Forest Degradation (REDD) is a scheme being promoted by the United Nations, the World Bank, Nicholas Stern, and various NGOs so as to protect intact forests, largely occupied by indigenous and peasant peoples. Corporations can earn carbon credits by investing in tree plantations and biofuel production. According to Lohmann (2009:1068), competition for the large sums of money has "already divided various indigenous peoples' groups and local communities, some of whom see REDD as an unprecedented opportunity for advancement, whilst others see it as a potentially catastrophic enclosure movement and violation of the sacred; and environmentalists who divide between the proponents such as the US Conservation International and the Nature Conservancy on the one hand and, on the other, groups such as FERN and the Forest Peoples Programme, who, looking to the example of the ill-fated Tropical Forest Action Plan of the 1980s and 1990s, see REDD as disempowering forest peoples in favour of acquisitive corporations and officials with little experience of or incentive to understand local issues of forest conservation." In an agroforest project in Guatemala, many of the indigenous subsistence farmers in the project area were pushed to the edge of the agricultural frontier as land in the fertile lowlands became concentrated in agribusiness sectors (Lohmann 2006:222). Besides the issue of displacement of indigenous and peasant peoples, one problem with forestation as a carbon mitigation strategy is the difficulty in measuring the amount of carbon sequestered thereby (Drake 2000:211).

Carbon offsets, whereby frequent flyers and big consumers can plant trees in developing societies, have often been compared to the purchase of indulgences by sinners during the medieval era to buy their way into heaven. According to Frank (2009:35),

> Natural forests, bogs, wetlands and grasslands, on which indigenous people must subsist, are foolishly being destroyed for this purpose after they have been conveniently declared degraded. The carbon traders have been cheating on calculations for the offsets, grossly overestimating the ability of the trees to absorb more carbon.

US-based Applied Energy Services reportedly created the first carbon-offset program when the company managed to have its plan to construct a 183-megawatt coal-fired power plant approved partly due to its promise to plant 50 million trees in the Guatemalan highlands (Smith 2007:14). In 1996 Future Forests was launched and attracted celebrities such as the Rolling Stones, Brad Pitt, and Jake Gyllenhaal as its clients. Other notable figures who participate in the voluntary CO_2 offset industry include Al Gore, Hillary Clinton, Arnold Schwarzenegger, and David Cameron (Rogers 2010:152).

Future Forests became the Carbon Neutral Company in 2005 and added renewable-energy and energy-efficiency projects to its portfolio (Smith 2007:17). Carbon Neutral Company has projects in Mexico, Mozambique, India, Uganda, and Tanzania. Another offset company, Climate Care, provides carbon offsets to British Airways passengers and operates a forestry project in Uganda (Smith 2007:25). Terrapass is a US-based offset company that encourages its subscribers to believe that "me and my car are doing something good for the planet" (quoted in Smith 2007:11). BP promotes "Global Choice" in Australia, which enables drivers to purchase BP Ultimate, a sulfur-free petrol, promising that "BP will automatically offset 100 percent of your emissions at no extra charge to you" (quoted in Smith 2007:12).

Kevin Smith (2007:13) provides a cogent critique of carbon-offset schemes:

> From flights, to four-wheel drives, to petrol itself, carbon offsets provide a false legitimacy to some of the most inherently unsustainable products and services on the market. What's more, the costs of the purchasable legitimacy are often largely shunted onto the consumer, who effectively ends up paying for the greenwash. These companies also benefit because offset schemes place more of the focus on the consumers' responsibility for climate change—at the expense of examining the larger, systemic changes that we need to bring about in our industries and economies.

Ultimately most emissions offset schemes shift the onus of responsibility from developed societies to developing ones. Indeed, some people have argued that "offsetting is like paying someone else to be faithful so that you can have an affair" (Simms 2009:234).

The Green New Deal

The United Nations Environment Programme (2009) and various other parties have called for a "Green New Deal." In his version of the Green New Deal, Andrew Simms (2009: 265), a British environmental economist, delineates three key components:

- Stabilization of the financial system that would dissolve "discredited financial institutions that have survived only through the injection of vast sums of public money"
- Investment in environmentally sustainable energy, transport, and building infrastructure
- Creation of a "low-carbon, high-wellbeing economy"

In another version of the Green New Deal, Tim Jackson (2009:7), economics commissioner of the Sustainable Development Commission in the United Kingdom, argues that the Global Financial Crisis of 2008 offers humanity a "unique opportunity to address financial and ecological sustainability together" by questioning the "underlying vision of prosperity built on continual growth." In keeping with the Jevons paradox, he acknowledges that improvements in energy and carbon intensity tend to be offset by increased economic growth. Rather than adopting a simple Keynesian stimulus to return the global economy to former economic growth patterns, he borrows from Herman Daly's notion of a steady state economy (Jackson 2009:77). Jackson calls upon governments to invest in public infrastructure, reduce social inequality, redistribute existing jobs and reduce work hours, reverse the culture of consumption, implement resource/emissions caps, and shift to alternative energy sources that will help to stabilize CO_2 emissions. He argues that a "macro-economy predicated on continual expansion of debt-driven materialistic consumption is unsustainable ecologically, problematic socially, and unstable economically" (Jackson

2009:103). As a green Keynesian economist, Jackson appears to assume that global capitalism can function as a nongrowth system when history repeatedly has told us that, by its very nature, it must grow or die out. As Foster (2011:29) maintains,

> A society based on economic contraction cannot exist under capitalism. . . . Getting rid of capitalists and banning wage labor, currency, and private ownership of the means of production would plunge society into chaos. It would bring large-scale terrorism. . . . We need to find another way out of development, economism (a belief in the primacy of economic causes and factors) and growth: one that does not mean forsaking the social institutions that have been annexed by the economy (currency, markets, even wages) but reframes them according to different principles.

In chapter 6, I explore the creation of an alternative world system that operates according to such different principles.

The United Nations Environment Programme convened a consultative meeting of policy experts in Geneva on December 2 and 3, 2008, under the auspices of the Green Economy Initiative to outline possible proposals for a Global Green New Deal (GGND) and commissioned Edward B. Barbier (2010), an economics professor at the University of Wyoming, to delineate the key components of the scheme. He identifies three principal aims of any GGND as consisting of the following:

- Revival of the global economy by creating employment opportunities, including green jobs, and protecting vulnerable groups
- Reduction of carbon dependency, ecosystem degradation, and water scarcity
- Advancing the Millennium Development Goal of ending extreme world poverty by 2025 (Barbier 2010:xix)

As table 5.2 indicates some G20 countries have included "green" investments in their stimulus packages, which were created in response to the Global Financial Crisis. According to Barbier (2010:19),

Table 5.2. Top 10 Countries by Green Stimulus Spending as Percentage of Total Fiscal Expenditures

Country	*Green Stimulus as Share of Total Fiscal Stimulus*
United Kingdom	10.6
South Africa	10.7
United States	12.0
Germany	13.2
Australia	21.2
France	21.2
Norway	31.0
China	33.4
European Union	58.7
South Korea	95.2
Global share	**15.4**

Source: Adapted from Barbier (2010:21).

> South Korea's Green New Deal plan is equivalent to around 3 percent of GDP and is expected to create 960,000 new jobs by 2012. Over a third of China's stimulus spending is for energy efficiency and environmental improvements, rail transport and new electricity grid infrastructure, while the UK government has devoted around 11 percent of its fiscal stimulus to green investments and aims to create 400,000 new jobs over the next eight years.

Ultimately, green capitalism fails to address the treadmill of production and consumption that contributes to greenhouse gas emissions and ultimately climate change, and it does not address the issue of exporting polluting industries from developed to developing countries. Green capitalism tends to be oblivious to social justice issues, such as growing social inequality, or at best tends to downplay them or pay them lip service.

Ecological Modernization

Ecological modernization has become a virtually hegemonic stance that asserts that environmental sustainability and effective climate change mitigation can be implemented by adopting more efficient, environmentally friendly, and low carbon-emitting energy sources and manufacturing processes. Carter (2007:228) lays out the key components of ecological modernization as follows:

> Ecological criteria must be built into the production process. On the supply side, costs can be reduced by improving productive efficiency in ways that have environmental benefits. Savings can be made by straightforward technological fixes to reduce waste, and hence pollution, but also through a more fundamental rethinking of manufacturing processes so that large-scale systems such as "smoke-stack" industries, that can never been made ecologically sound, are gradually phased out. On the demand side, there are growing markets in green technologies such as air pollution abatement equipment and alternative forms of energy. The rise of "green consumerism" has stimulated demand for goods that minimise environmental damage both in the way they are made (by using recycled materials or minimising packaging) and in their impact when used (by containing less harmful chemicals such as phosphate-free washing powders).

Lester R. Brown, the author of *Plan B* and director of the Earth Policy Institute, is a staunch proponent of ecological modernization. Steve Pascala and Robert Socolow present an ecological modernization scheme of what they term *stabilization wedges*, which include the following options:

- Energy efficiency and conservation: efficient vehicles, reduced use of vehicles, efficient buildings, and efficient coal plants
- A fuel shift from coal to natural gas
- CO_2 capture and storage (CCS)
- Nuclear power
- Renewable electricity and fuels
- Forests and agricultural soils (see Maslin 2009:150)

While Al Gore in his *An Inconvenient Truth* (2006) popularized the findings of climate science among millions of people around the world, his solutions are framed very much within the parameters of green capitalism and ecological modernization by advocating carbon trading, green consumerism, tree plantations, and techno-fixes as sufficient climate change mitigation strategies. He proposes implementation of a Global Marshall Plan, which would entail the following elements: (1) stabilization of the world population, (2) the development and sharing of "appropriate technologies," and (3) the development of a "new global eco-nomics" (Gore 2007:307–37). Gore (2007:346) argues that the definition of GNP should be changed to include environmental costs and benefits, or treated as what mainstream environmental economists term an *externality*.

In his most recent book, *Our Choice: A Plan to Solve the Climate Crisis* (2009), Gore lays out his views on ecological modernization, which include an overall endorsement of energy efficiency, retrofitted buildings, hybrid cars, greater reliance on public transport, and renewable sources of energy (solar, wind, and geothermal) as climate change mitigation strategies. Overall, he is ambivalent about the viability of carbon capture and sequestration (Gore 2009:134–49), at least for the immediate future, noting, "Most experts who have studied the CCS option have concluded that it is probably impracticable for many years to come, because the technology for capturing CO_2 would either require a dramatic increase in the use of coal and gas for the same amount of electricity, or sharply reduce the amount of electricity obtained from burning the same amount of fuel as of present—and because every one of the potential geological repositories presents a unique and extremely difficult challenge in characterizing its geology deep underground and estimating both storage capacity and the safety of storing CO_2 there." He also is highly ambivalent about the viability of the "nuclear option" (Gore 2009:150–69) for a variety of reasons, including the potential for nuclear power plants to have to shut down during heat waves, as was the case during the 2003 European heat wave, or their heavy reliance on water in increasingly drought-stricken areas. While Gore (2009:123) expresses concern about producing ethanol for motor vehicles from corn and even sugar, he argues that "the second generation technology for producing ethanol—when it becomes commercially available—has a significant advantage over the first generation technology; instead of using food crops, it will make liquid fuels from perennial grasses, fast-growing trees, and waste stream with a high cellulose content."

Some oil companies have started to move into the renewable energy source sector, albeit in a very modest way. According to Newell and Paterson (2010:44),

> In May 1997, BP's chief executive officer (CEO) John Browne decided that there was mileage in being seen to be green, announcing the shift of strategy at a high-profile talk at Stanford University. With his backing, the company re-branded itself "Beyond Petroleum," an ambitious claim for a company whose 2005 accounts indicate that the company invested just $800 million a year into its "Alternative Energy" division, representing just 5.7 percent of its 2005 total capital investment, while 72 percent of BP's new capital investment was spent looking for more oil and gas. . . . The current financial crisis seems to have damaged BP's alternative energy budget, which was down from $1.4 billion in 2009, while in April of the same year the company closed a number of solar-panel manufacturing plants in Spain. At the same time the company is increasing investments in controversial oil sands extraction in Alberta, Canada.

In 2010, BP's efforts to find even more oil in the Gulf of Mexico resulted in the largest oil spill on record. In contrast to BP with its efforts to look green, Shell withdrew itself from investment in wind, solar, and hydropower on the grounds that these endeavors would not prove sufficiently profitable (Newell and Paterson 2010:45).

Renewable Energy Sources, Energy Efficiency, and Hybrid Cars

Plan B advocates hope to replace nonrenewable energy sources with renewable ones. Advocates of *Plan B* include Al Gore, various nature conservation and wilderness societies, and many environmentalists, including ones in the climate movement (Murphy 2008:113). The "sunrise industries" that are touting energy conservation, efficiency, and renewable energy resources have "been working with some environmental groups to make the case for tough targets to stimulate markets for their products (for example groups like E7 and the European Wind Energy Association)" (Newell and Paterson 2010:42). Renewable energy generators will require equipment and buildings that will have to be produced by manufacturing processes that require fossil fuels and mineral resources. While I firmly believe that renewable energy sources, including solar, wind, and geothermal, have the potential of being part of the process of mitigating climate change, in and of themselves they are not a panacea. As Wall (2010:11) observes, "Even a renewable energy–fuelled capitalism would still tend to degrade the environment through commodification of nature."

Carbon Capture and Storage, Natural Gas, Nuclear Energy, Biofuels, and Biochar

As Black (2006:278) observes, "For many, the concept of clean coal is oxymoronic, like natural hairspray." Carbon sequestration, or carbon capture and storage, is as yet an unproven technology that supposedly will capture CO_2 and inject it into the ground, or even into the ocean, or possibly store it in old oil reservoirs or coal mines. If ever achieved on a mass scale, it would probably increase the cost of electricity generation and reduce energy efficiency in that it would require additional energy. Again, assuming the CCS technology is ever perfected, according to Murphy (2008:87),

> the potential amounts to be buried or sequestered are unbelievably large. There will be tens of thousands of power generators around the world, each providing a steady stream of CO_2 twenty-four hours a day, seven days a week, 52 weeks per year, feeding into a huge infrastructure of trains, pipes and ships delivering the gas to places where it will be pumped into the earth.

Finally, there may not be a sufficient number of leakproof sites around the globe to store large amounts of ever-increasing carbon. With respect to the now biggest emitters of CO_2, whereas there are many deep saline aquifers in the United States, there are few in China (Richter 2010:91).

The emerging global natural gas market, which is heavily dependent on the liquefied natural gas industry, offers an "example of a corporate-endorsed solution to the simultaneous ecological and economic 'crises' associated with fossil capitalism" (Zalik 2008:41). Liquid natural gas (LNG) is exported by countries with large natural gas reserves, such as Algeria, Australia, Brunei, Indonesia, Libya, Malaysia, Nigeria, Oman, and Trinidad and Tobago. Around 60 LNG receiving terminals are located in primarily developed and more advanced developing societies, such as the United States, Japan, various European countries, and South Korea.

Nuclear energy is frequently cited as a clean form of energy. There are presently some 439 nuclear reactors in operation around the world, 104 of them located in the United States (Charman 2008:39). France, with 59 reactors, obtains 76.9 percent of its electricity from nuclear power, and Japan, with 56 reactors, obtains 27.5 percent of its electricity from this source, deriving much of the rest from hydroelectric plants. The United States, Canada, the United Kingdom, Italy, China, India, and South Africa all have plans to construct new nuclear reactors. While such plans do not presently exist in Australia, various prominent individuals in that country, including popular science writer Tim Flannery and Barry Brook of the University of Adelaide, are staunch proponents of nuclear power. Nuclear power is incredibly expensive, with a new 1,600-megawatt reactor costing about US$6 to 7 billion (Charman 2008:41). Furthermore, the construction of nuclear power plants requires large amounts of fossil fuel, which results in greenhouse gas emissions. The Energy Watch Group (2006) estimates that the world's proven and obtainable uranium resources would only last between 30 and 70 years. As for the increasingly voiced assertions that nuclear power plants will serve as a climate change mitigation strategy, Jim Green (n.d.:3) argues,

> Claims that nuclear is "greenhouse free" are incorrect as substantial greenhouse gas emissions are generated across the

> nuclear fuel cycle. Fossil-fuel generated electricity is more greenhouse intensive than nuclear power, but this comparative benefit will be eroded as higher-grade uranium ores are depleted. Most of the earth's uranium is found in very poor grade ores, and recovery of uranium from those ores is likely to be considerably more greenhouse intensive.

Helen Caldicott (2009), an Australian pediatrician and longtime antinuclear activist, not only discusses the lethal dangers of nuclear power but also asserts that it contributes to climate change. She maintains that "nuclear power is not 'clean and green,' as the industry claims, because large amounts of traditional fossil fuels are required to construct the massive concrete reactor buildings, and to transport and store the toxic radioactive wastes created by the nuclear process" (Caldicott 2006:xiii). Furthermore, nuclear reactors are dangerous, as the incidents at Three Mile Island in Pennsylvania and Chernobyl in the Ukraine revealed. Nuclear reactors leak radioactivity into adjacent groundwater and soil (Smith 2006). Finally, nuclear power plants produce plutonium which can be used to develop nuclear bombs.

Fears of the risks of nuclear power reactors came to fruition with the Fukushima nuclear disaster in the wake of the earthquake and tsunami in Japan on March 11, 2011. Despite the fact that the nuclear disaster will have serious impacts on health and economic and social life, as Barbara Rose Johnston (2011:3) observes, "what is most disturbing for Japanese citizens, residents, and to a much lesser degree for those downwind, is that—despite the acknowledgments that the worst has happened, that emissions continue to persist, that there is no viable plan for the safe control and cleanup—the average citizen still struggles to access current and meaningful information that might inform proactive action that reduces exposure and minimizes risk."

David MacKay (2009), a physics professor at Cambridge University, argues that unless sustainable energy sources cover a large area, their contribution to energy supply will be very small, and nuclear power will need to be added to the energy mix. James Hansen (2009) advocates the rapid development of alternative energy sources, including fourth-generation nuclear power that would rely upon fast-breeder reactors, as well as a carbon tax, the phasing out of coal, and reforestation as climate change mitigation strategies. Fourth-generation nuclear reactors would theoretically burn up nuclear waste while generating power; unfortunately, however, they are still in the experimental stage. According to MacKay (2009:163),

> Uranium can be used 60 times more efficiently in fast breeder reactors, which burn up all the uranium—both the 238U and the 235U (in contrast to the once-through reactors, which burn mainly 235U). As long as we don't chuck away the spent fuel that is spat out by once-through reactors, this source of depleted uranium could be used too, so uranium that is put in once-through reactors need not be wasted.

Fast-breeder reactors may not be available for decades, they would be expensive to construct, and their safety or lack thereof has not been determined (Li 2009:1047). Nuclear fusion technology, while theoretically safer than nuclear fission technology, remains unavailable for decades.

Biofuels often are touted as an alternative form of energy. Actually, they exist in four forms: (1) wood products and crop residues that can be burned; (2) ethanol derived from sugars, starches, and cellulose; (3) biodiesel derived from oil crops or waste cooking oil; and (4) methane derived from natural gas, animal manures, and human sewage. In the United States alone, the price of corn has skyrocketed as a result of ethanol production. ADM, a $44 billion a year company, is one of the largest producers of biofuels and has been heavily subsidized by the US government. To obtain large government subsidies, "ADM's chairman, Dwayne Andreas, contributed huge sums of money to both Republicans and Democrats from the Nixon administration through the Clinton years" (Magdoff 2008:41). Biodiesel, which is a more common biofuel in Europe than in the United States, can be produced from soybeans, oil palms, and rape or canola. Unfortunately, the production of biofuels requires huge amounts of farmland largely to fuel motor vehicles. Indeed, as Fred Magdoff (2008:42) asserts, the "use of large quantities of potential food crops—especially corn (maize) and soybeans, but also including such crops as oil palm—to produce fuels is a major contributing factor to the current world food crisis." Furthermore, the production of corn ethanol requires massive resources, including diesel fuel for farm machinery, chemical fertilizers and pesticides, and large amounts of irrigation water (Frank 2009:38). Huge agribusinesses have come to dominate biofuel production. For example, in the United States, Cargill, ADM, and ConAgra control over 80 percent of US corn exports; Cargill is the largest exporter of raw sugar from Brazil; and Wilmar, IOI, Synergy Drive, and Cargill dominate 60 percent of the palm oil trade (Shiva 2008:83). The demand for biofuels has increased the price of agricultural products in many parts of the world. On July 3, 2008, the *Guardian* "claimed that US and EU agrofuel policies were responsible for three quarters of the 140 percent increase in food prices between 2002 and February 2008" (Bello 2009:105).

Biochar, a fine-grained, porous charcoal that is resilient to decomposition, has been used by indigenous peoples as a means of restoring carbon to depleted soils and has the potential to sequester large amounts of CO_2. In addition to its naturally occurring form, biochar can be manufactured by burning wood, switch grass, manure, or other types of biomass. While possibly having some merits, biochar is not a panacea as a climate change mitigation strategy because, as Lohmann (2009:1068) argues, like biofuels, it "would involve altering land-use practices over millions of hectares in untried ways." Indeed, the Kuna people of Central America refer to biochar as "bioshit."

Geo-Engineering

Various parties have proposed a wide range of geo-engineering remedies, or what might be called *climate engineering*, as climate change mitigation strategies. These include placing a gigantic sunshade in space so as to drastically cool the Earth's average temperature; injection of sulfate aerosols into the atmosphere to increase cloud cover, thus increasing solar reflectivity; removal of CO_2 from the atmosphere by photosynthesis by growing ocean plants, such as plankton; and enrichment of some ocean areas with iron to better absorb CO_2. Edward Teller, a renowned nuclear physicist, has proposed placing "billions of tinfoil strips in orbit around the Earth in order to reflect up to 2 percent of the incoming sunlight and cool down the planet" (Gore 2009:314). Aside from the expense involved, a pitfall of many geo-engineering schemes is that their potential negative consequences have not been determined.

Although proponents of ecological modernization concede that many environmental problems are by-products of a market economy or global capitalism, they generally reject transcending the capitalist mode of production. Proponents of ecological modernization maintain that capitalism can be made more "environmentally friendly" through environmental regulations and technological changes managed by ecologically sensitive governments, or *green states*, that function in concert with corporations (Clark and York 2005). Ecological modernizations, which started in northern European countries, have quickly been adopted by corporate elites and politicians in various countries, including the United States and Australia, and even among mainstream environmental groups in both developed and developing societies. While adopting more environmentally sustainable technologies and achieving energy efficiency are in and of themselves commendable objectives, they will not lead to a "decoupling" from economic growth as mainstream environmental economists maintain because, following the Jevons paradox, in a capitalist economic system, "energy savings are used to promote new capital formation and the proliferation of commodities, demanding ever greater resources" (Foster, Clark, and York 2010:5). Salleh (2010:196) notes that ecological modernization "will consume vast amounts of front-end fuels—in, welding turbines and grids, road making, water supply, component manufacture for housing, air conditioning for shopping malls."

The Climate Denialists

Although many corporations, mainstream economists, and others have come to embrace some variant of green capitalism and/or ecological modernization, many corporations, neoconservative think tanks, and conservative pundits fall into a disparate category termed *climate skeptics*, or what I prefer to term *climate denialists*. Within this amorphous grouping are several categories of bodies or people: (1) those who deny the reality of climate change; (2) those who admit its existence but claim that anthropogenic forces are not primarily responsible for it, thus attributing it primarily to natural forces; and (3) those who admit that anthropogenic activities are partly responsible for climate change but deny that it matters or argue that various social benefits may accrue from it. Morgan and McCrystal (2009:229) argue that acceptance of mainstream climate science is closely related to an individual's political perspective. They assert that the "conservative end of the spectrum tends to regard the whole theory of anthropogenic global warming as a stalking horse for anti-consumerist Greenies and interventionist liberals, and to regard any attempt to introduce policies to curb greenhouse emissions as an unconscionable interference with the workings of the 'free market'" (Morgan and McCrystal 2009:229). They further assert, "Greenies and liberals, on the other hand, based on the nefarious tactics of some deniers, tend to regard all climate change scepticism as nothing more than a rearguard action by hard-pressed corporations whose interests are mortgaged to the fossil fuel economy" (Morgan and McCrystal 2009:229).

The Union of Concerned Scientists has conducted an analysis of contrarian groups and claims that the "great majority either belong to or are actively sponsored by organisations such as the fossil fuels industries" (Hillman 2004:22). For example, Patrick J. Michaels, a leading contrarian climate scientist, receives funding from the Western Fuels Association and edited the *World Climate Report* (McCright and Dunlop 2000:508). He has also had close links with the George C. Marshall Institute, a strong advocate for the Ronald Reagan administration's Star Wars initiative (Pearce 2010:82). Fred Singer, a politically ultraconservative climate scientist based for many years at the University of Virginia, some of whose research had been funded by the now defunct Global Climate Coalition, has acted over the years as one of the world's leading climate denialists. As Eric Pooley (2010:36) states,

> In the 1980s he disputed the link between the industrial chemicals called CFCs and ozone depletion. . . . In the 1990s he disputed the link between secondhand tobacco smoke and cancer. And for two decades he had been disputing the idea that greenhouse gases were warming the planet.

Although Singer finally admitted that climate change was real, he ascribed it to solar activity in his book *Unstoppable Global Warming: Every 1,500 Years* (2007). In contrast, Patrick J. Michaels, another University of Virginia climate scientist who has worked at the Cato Institute, has admitted that climate change is anthropogenic but says it is not dangerous. According to Vanderheiden (2008:xv), "Some climate skeptics, including the novelist Michael Crichton, who was called as the lead witness in a 2005 Senate Environment and Public Works Committee hearing on climate change, see an elaborate hoax perpetrated by environmental groups and dogmatic scientists designed to make the United States submit to an insidious world government."

A few Australian physical scientists with impressive credentials are in the contrarian camp. William Kinmouth, who served between 1986 and 1998 as the head of Australia's National Climate Centre, following the release of the IPCC Third Assessment Report in 2001, rejected the IPCC's conclusions that in large part recent climate change is due to anthropogenic causes. In *Climate Change: A Natural Hazard,* he faults the IPCC climate modeling system (Kinmouth 2004). More recently, Ian Plimer (2009), a geology professor at the University of Adelaide, argued in *Heaven + Earth* that climate changes are driven primarily by the Earth's position in the solar system, the sun, wobbles in the Earth's orbit, ocean currents, and plate tectonics—not anthropogenic activities that result in CO_2 emissions.

Contrarian organizations include the now defunct Global Climate Coalition and the Greening Earth Society (US), the Heritage Foundation (US), the Cato Institute (US), the Heartland Institute (US), the Competitive Enterprise Institute (US), the

George C. Marshall Institute (US), the Copenhagen Consensus (Denmark), the Fraser Institute (Canada), the Lavoisier Group (Australia), and the Institute of Public Affairs (Australia) (Beder 2004:27; Flannery 2005:243–45; Craven 2009:130–34).

The Global Climate Coalition was an energy industry lobby group that operated out of the Washington, DC, office of the National Association of Manufacturers (Linden 2006:273). Along with a small number of contrarian scientists, it maintained a frontal attack on IPCC conclusions. The coalition included the American Highway Users Alliance, the American Petroleum Institute, the Edison Electric Institute, the National Association of Manufacturers, and the National Mining Association (Gelbspan 2004:40); it was abandoned in 1999 by British Petroleum and Shell due to concern that membership would damage their public image. Indeed, British Petroleum renamed itself "Beyond Petroleum" to symbolize its willingness to invest in renewable energy sources. The coalition quickly was abandoned by Ford, United Technologies, Daimler-Chrysler, and the Southern Company. Some of the abandoning companies joined the Pew Center's Business Environmental Leadership Council. Nevertheless, as Newell and Paterson (2010:37) assert, "though [it is] now disbanded, it is difficult to overestimate the importance of the GCC during the early to mid 1990s as *the* voice of concerned industry in the international climate negotiations." In addition to contrarian think tanks, scientists, and influential individuals, various corporate-based organizations have either supported these actors or opposed policies, such as a proposed European Union carbon tax in 1992, that threaten their interests. Such groups have "included the Confederation of British Industry (UK), the World Coal Institute, the American Petroleum Institute or Western Fuels Association (US), as well as regional groupings such as employers' organisations like the Union of Industrial Employers' Confederations in Europe (UNICE, now called Business Europe) and the European Round Table of Industrialists" (Newell and Paterson 2010:38).

Bjorn Lomborg, a Danish statistician self-billed as a "skeptical environmentalist," has engaged in an astute international campaign to convince the mass media and various public policy makers that the seriousness of climate change has been exaggerated and that it is not as serious as other global problems, such as poverty and HIV/AIDS (Friel 2010). He has even managed to convince the highly prestigious Cambridge University Press to publish his two major books (Lomborg 2010, 2007). Lomborg established the Copenhagen Consensus Center, which has commissioned reports seeking to evaluate various geo-engineering schemes as effective and relatively inexpensive ways of reducing greenhouse gas emissions (Perlmutter and Rothstein 2011:183).

While various progressive Christian clerics have become environmentalists of sorts by arguing that humans must act as stewards of a fragile planet, the ultraconservative archbishop of Sydney, Australia, Cardinal George Pell, maintains that environmentalists manifest a new "pagan emptiness" and made the following critique of climate science in January 2008:

> The public generally seem to have embraced even the wilder claims about man-made climate change as if they constituted a new religion. These days, for any public figure to question the basis of what amounts to a green fundamentalist faith is tantamount to heresy. (Quoted in Flannery 2010:38)

Conclusion

Newell and Paterson (2010) delineate four possible future scenarios for climate capitalism. The first of these they term a *climate capitalist utopia*, which will entail a rapid decarbonization of the global economy and "investment in renewable energy, energy efficiency and conservation, carbon capture and storage, advanced public transport and urban infrastructure reform," along with emissions trading (Newell and Paterson 2010:165). The second scenario would be *stagnation*, where carbon markets fail, and international efforts to set emissions targets collapse, resulting in a dystopian situation in which humanity seeks at best to "adapt to whatever climate change has to offer" (Newell and Paterson 2010:168). Scenario three would be a *decarbonized dystopia* where a low-carbon global economy is achieved but engages in a type of carbon colonialism in which "money pours into biofuels both in the North and South, producing large mono-crop plantations with appalling working conditions, the destruction of biodiversity, and price rises of key food crops which place them beyond the reach of the poor" (Newell and Paterson 2010:169–70). In addition, climate refugees will be denied access to less climate-ravaged countries or held in camps indefinitely (Newell and Paterson 2010:172). Newell and Paterson (2010:178) deem their fourth scenario, *climate Keynesianism*, the most favorable in that governments would regulate carbon markets and implement redistributive mechanisms both within and between nation-states and thus "create stable conditions for investment in carbon markets and in renewable energy, energy efficiency, and so on." Ultimately, they anticipate that over the course of the next 20 to 30 years, the most likely outcome will be "some messy mix" of their four scenarios, with "some areas of the world stagnating, others going ahead with a pure neoliberal version, while others still regulate the carbon economy even more stringently" (Newell and Paterson 2010:178).

Tadzio Mueller and Alexis Passadakis (2010:562–64) delineate eight theses against green capitalism. I list a condensation of their first four theses:

- "Green capitalism will not challenge the power of those who actually produce most greenhouse gases."
- "All types of green capitalism fail to acknowledge that the expansive nature of capitalism—its need to grow—will undermine any attempt to reduce its constant imperial demand for more resources."
- "In a green capitalist setup, wages will probably stagnate or even decline, to offset the rising costs of 'ecological modernization.'"
- "The 'green capitalist state' will be an authoritarian one."

Ultimately, Mueller and Passadakis (2010:563) insist that governments and corporations will not provide adequate solutions to the climate crisis and that, rather, the solutions will have to emerge from "globally networked social movements for climate justice," a topic that I explore in chapter 6.

It is important to note that some components of ecological modernization—such as renewable sources of energy (solar, wind, and geothermal), improved energy efficiency and building construction and design, and a massive shift from private vehicles to energy-efficient public transport systems—have the potential to serve as important climate change mitigation strategies. However, as Hornborg (2001:25–26) persuasively argues,

> What ecological modernization has achieved is a neutralization of the formerly widespread intuition that industrial capitalism is at odds with global ecology. . . . The discursive shift since the 1970s has been geared to disengaging concerns about environment and development from the criticism of industrial capitalism as such. But the central question about capitalism should be the same now as it was in the days of Marx: Is the growth of capital of benefit to everybody, or only to a few at the expense of others?

Indeed, the facts repeatedly speak for themselves in answering a resounding yes to the latter half of that question. Ultimately, technological innovations that on the surface appear to be more environmentally sustainable and energy efficient, as I argue later in this book, must be part and parcel of a shift to a steady-steady or zero-growth global economy if they are to circumvent the Jevons paradox associated with global capitalism and its need for constant economic growth.

6

A Vision of an Alternative World System: Toward Global Democracy Based on Social Justice and Environmental Sustainability

British sociologist Anthony Giddens (2009:50) somewhat reluctantly admits that there is a "left/right tinge to current climate change debates: those who want to respond to climate change through wide-spread reform mostly tend towards the political left; most of the authors who doubt that climate change is caused by human agency, on the other hand, are on the right." In reality, there are many positions on the left as to what sorts of policies and actions will be necessary to contain dangerous climate change. Many people on the moderate left argue that effective climate change mitigation can be successfully implemented by regulating and modernizing global capitalism, whereas those on the far left argue that climate change mitigation ultimately must entail transcending global capitalism and replacing it with a world system committed to social parity, democratic processes, and environmental sustainability. While Hulme (2009:362) is hardly a radical, let alone an eco-socialist, I am wholeheartedly in agreement with his assertion that "we need to see how we can use the idea of climate change—the matrix of ecological functions, power relationships, cultural discourses and material flows that climate change reveals—to rethink how we take forward our political, social, economic and personal projects over the decades to come." Thus, climate change compels us to engage in what Immanuel Wallerstein (1998:1) terms *utopistics,* which he defines as the "serious assessment of historical alternatives, the exercise of our judgment as to the substantive rationality of alternative possible historical systems." In seeking to assess possible future scenarios with respect to climate change, one must consider the possibility of a dystopian future with the hope that this will contribute to the realization that serious mitigation efforts will require an alternative to global capitalism, one that is based on both social equity and environmental sustainability and that will allow humanity to reach a steady state for itself and other forms of biological life, both large and small.

The Road to Dystopia

Climate change scenarios prompt us to imagine dystopian visions of the future, if for no other reason than to forewarn us to take serious measures to counteract possible doomsday events. In this section I discuss possible dystopian scenarios resulting from climate change as depicted by journalist Mark Lynas, James Lovelock, scholars calling for eco-authoritarian climate regimes, security analysts, and neoliberal analysts who divide the world into winners and losers in terms of the impact of climate change on human populations or regions.

Mark Lynas

In his book *Six Degrees*, journalist Mark Lynas (2007), based on his perusal of numerous climate scientific reports, vividly portrays climate change scenarios at 1°C to 6°C increases in the global temperature, most of which will have negative impacts on human populations.

One Degree

- Perennial drought in the western United States, resulting in the devastation of agriculture
- Possible higher rainfall in the US Midwest, accompanied by increased agricultural productivity

- Possible decrease of temperatures in northwestern Europe because of a slowdown of the Gulf Stream and North Atlantic Conveyor Belt
- Disappearance of snow on Mount Kilimanjaro
- Increased rainfall in the Sahel
- Parching of the Amazon Basin
- Beginning of an Arctic meltdown
- Thawing of mountain slopes in the Alps
- Serious bleaching of the Great Barrier Reef

Two Degrees

- Serious flooding in southern China
- Acidification of the oceans
- Droughts in the Mediterranean area
- Opening up of shipping routes across an ice-free Arctic Ocean
- Almost complete disappearance of the Arctic tundra and northward march of the taiga
- Formation of huge glacial lakes in the Himalayas
- Disappearance of glaciers in the Cordillera Central of the Andes
- Increased wheat and maize production in new areas of western Russia and southern Scandinavia
- Reduced fish populations on both sides of the Atlantic

Three Degrees

- Serious drought in Botswana and southern Africa, accompanied by fierce dust storms in the Kalahari Desert
- Possible shift to a "super El Niño" accompanied by massive floods and mud slides in California
- Parching and death of the Amazon Rainforest
- Reversal of the carbon cycle, with vegetation and soil starting to release CO_2
- Increased drought and bushfires in most of Australia, making life on much of the continent impossible

- Frequent and intense hurricanes along the Gulf Coast of the United States
- Devastating monsoons on the west coast of India and in the Bay of Bengal, Bangladesh, and northeastern India
- Loss of massive quantities of water in Pakistan
- The submergence of parts of New York City

Four Degrees

- Loss of one-third of Bangladesh's land area, resulting in the displacement of millions from the Meghna Delta
- Endangering by flooding of low-lying and deltaic cities such as Shanghai, Mumbai, Alexandria, Boston, New York, New Orleans, London, and Venice
- Massive shrinking of Greenland's ice sheet into center of landmass
- Slowing and shutdown of the North Atlantic Conveyor Belt
- Spreading of new deserts in southern Europe
- Possible July and August temperatures of 48°C in Switzerland, accompanied by wildfires and diminished water supplies
- Completely ice-free summer in the North Pole
- Release of CO_2 contained in frozen Arctic soils

Five Degrees

- Remaining ice sheets eliminated from both poles
- Rising sea levels inundating coastal cities and penetrating far inland
- A 20 percent decrease in Nile flows
- Disappearance of nearly 90 percent of California's snowpack
- Human populations greatly restricted in terms of habitable areas due to drought and flooding
- Northern Europe potentially becoming a crowded refugee area
- Patagonia, Tierra del Fuego, Tasmania, the South Island of New Zealand, and the ice-free Antarctic Peninsula becoming other potential refugee areas

Six Degrees

- Small eruption of oceanic methane potentially causing mass extinction
- Possible ocean stratification and hydrogen sulfide poisoning
- Possible creation of artificial atmospheres or establishment of colonies on other planets

James Lovelock

James Lovelock, inventor of the Gaia hypothesis, asserts that overpopulation constitutes the roots of humanity's environmental problems. He recommends that humanity stabilize its population at 500 million to 1 billion and warns that Gaia will cull those who break the rules (Lovelock 2006:180–81). Lovelock (2006:171) argues that future society will be tribal and fractionated between the privileged and the poor. He states, "I think we have little option but to prepare for the worst and assume that we have already passed the threshold," and "we face unrestrained heat, and its consequences will be with us within no more than a few decades." (Lovelock 2006:196). Lovelock (2006:200) goes on to argue that "we can neither prepare against all possibilities, nor easily change our ways enough to stop breeding and polluting. Those who believe in the precautionary principle would have us give up, or greatly decrease, burning fossil fuel."

Given that the body politic is addicted to economic growth, he asserts that humanity must turn to fission nuclear energy to "keep the lights of civilization burning until clean and everlasting fusion is available" (Lovelock 2006:14). Lovelock also advises humanity to turn to various other climate change mitigation strategies, including high-density living, geo-engineered manipulation of the Earth's albedo, extensive tree planting, fertilization of ocean algal ecosystems with iron, carbon sequestration, synthesis of food from inorganic raw materials, and production of biofuels. He argues that alarmists have exaggerated the dangers of nuclear energy and is highly critical of "soft energy," particularly wind power, which he believes will mar the beauty of the countryside with numerous wind farms. Lovelock is very critical of urban society, despite the fact that he advocates high-density living, and he views rural dwellers such as himself as still in touch with nature.

In *The Vanishing Face of Gaia*, Lovelock (2009) identifies portions of the Earth that may be inhabitable in a dystopian future. These include the northern regions of the United States and Russia, Canada, Scandinavia, Siberia, Patagonia, southern Chile, and island nations or states, such as Japan, Tasmania, New Zealand, and the British Isles, along with many smaller islands, such as Hawaii, Taiwan, and the Philippines (Lovelock 2009:11). Lovelock (2009:57) refers to such places as "lifeboats for humanity" and grants that the various continents will have "oases and river courses still watered well enough for plants to grow" (Lovelock 2009:11). He predicts that the summer heat of Continental Europe will become increasingly unbearable, even with the use of air-conditioning (Lovelock 2009:61). Lovelock (2009:56) contends that a 4°C hotter planet may only be able to sustain a population of "as little as 100 million if the carrying capacity of the land surface of a hot Earth falls to 10 percent of what we have now."

Calls for Eco-Authoritarian Regimes

Perhaps frustrated by the cumbersome nature of global and national governance processes, including in liberal democracies, various scholars have argued that democratic processes are moving too slowly to contain climate change, and they suggest that eco-authoritarian, or even eco-fascist, regimes are needed to do so. James Anderson (2006:245) argues that the "radical changes necessary to sustain capitalism could indeed turn out to be an extremely authoritarian *counter-revolution*."

David Shearman and Joseph Wayne Smith (2007) maintain that "democratic states" are too dominated by special interest groups and materialism to create effective climate change mitigation policies. In reality, they assert, Western developed societies constitute plutocracies ruled by wealthy people (Shearman and Smith 2007:91). Following William Ophul in *Ecology and the Politics of Scarcity*, Shearman and Smith assert that liberal democracies need to be replaced by authoritarian states, such as Singapore, which will be governed by "natural elites" who have been socialized from childhood to address complex problems, such as climate change. They assert that humans are genetically wired to submit to authoritarian social structures (Shearman and Smith 2007:130). Shearman and Smith (2007:134) assert that climate change will create an economic and ecological disaster that will require a future government led by "specially trained philosopher/ecologists" committed to environmental sustainability.

In a somewhat similar vein, Lovelock (2009:61) maintains that "orderly survival . . . may require, as in war, the suspension of democratic government for the duration of the survival emergency." He goes on to argue,

> We have no option but to make the best of national cohesion and accept that war and the warlords are part of it. For island havens an effective defence force will be as important as our own immune system. Like it or not we may have to increase the size of and spending on our armed forces. Perhaps the next generation of scientists and engineers will be competent and serve the earth as general practitioners service us in medicine. (Lovelock 2009:62)

In a lifeboat world, Lovelock (2009:161) asserts that rules will have to be written to determine which climate refugees are granted a "safe haven in those few parts where the climate is tolerable and food is available."

In a similar vein, Lieven De Cauter (2008:111) suggests that climate change might contribute to a future world that "looks like some version of *Mad Max*, a trash sci-fi movie in which oil scarcity has turned the planet into a low-tech, chaotic, neomedieval society run by gangs." She argues that environmental disasters, such as Hurricane Katrina in New Orleans in 2005 and the tsunami in Indonesia in 2004, are contributing to the rise of what Naomi Klein has termed *disaster capitalism* under which the affluent sequester themselves from the victims of disasters in gated communities and green zones, or a "sort of security stronghold as well as an ecological safe haven" (De Cauter 2008:115).

The Potential for Climate Wars

Various defense agencies, including ones in the United States, the United Kingdom, and Australia, along with other organizations around the world, have made note of the "security risks" associated with climate change (see Webb 2007; Busby 2007; Council of the European Union 2008). In its recognition that global warming or climate change may pose a "security threat" to the United States, the Pentagon commissioned the CNA Corporation (n.d.), a nonprofit national security organization, to write a report on this issue. CNA convened a panel of retired military officers and national security experts as part of its effort to assess the security implications of global warming. In its report, CNA (n.d.) asserts that global warming "acts as a threat multiplier for instability in some of the most volatile regions of the world" and "will seriously exacerbate already marginal living standards in many Asian, African, and Middle Eastern nations, causing widespread political instability and the likelihood of failed states." CNA stresses that global warming poses the possibility of an even greater number of people attempting to emigrate, either legally or illegally, from Mexico to the United States, and more turbulent seas could adversely affect US naval operations in the North Atlantic.

Peter Schwartz and Doug Randall (2003:1) also authored another Pentagon-commissioned report titled "An Abrupt Climate Change Scenario and Its Implications for United States National Security," laying out worst-case scenarios that, "although not most likely," are "plausible" and thus would "challenge United States security in ways that should be considered immediately." While climate scientists may very well argue with their abrupt change scenarios, Schwartz and Randall (2003:2) envisage the possibility, over the next few decades, of an annual average temperature increase of up to 5°F (2.75°C) in Asia and North America and 6°F (3.3°C) in northern Europe; an annual average temperature increase of up to 4°F (2.2°C) in key areas of Australia, South America, and southern Africa; longtime drought in "critical agricultural resource regions for major populations in Europe and eastern North America"; and intense winter storms and winds in western Europe and the North Pacific region. Due to these climatic changes, they envisage the possibility of food shortages due to diminished "net global agricultural production," "decreased availability and quality of fresh water in key regions," and "disrupted access to energy supplies due to extensive sea ice and storminess." Schwartz and Randall (2003) lay out possible conflict scenarios for Europe, Asia, and the United States for 2010 to 2020 and 2020 to 2030. They envisage the possibility of conflicts in the Persian Gulf and Caspian Sea regions in 2020 due to increased oil prices; a civil war in China and border wars with adjacent countries in Southeast Asia; and also an "internal struggle in Saudi Arabia" in 2025, which "brings Chinese and U.S. naval forces to [the] Gulf in direct confrontation" (Schwartz and Randall 2003:17). According to Schwartz and Randall (2003:2),

> As global and local carrying capacities are reduced, tension could mount around the world, leading to two fundamental strategies: defensive and offensive. Nations [such as the United States and Australia] with the resources to do so may build virtual fortresses around their countries, preserving resources for themselves. Less fortunate nations, especially those with ancient enmities with their neighbours, may initiate in struggles for access to food, clean water, or energy.

More recently, the German Advisory Council on Global Change (2007:1) released a report stating,

> Climate change will draw ever-deeper lines of division and conflict in international relations, triggering numerous conflicts between and within countries over distribution of resources, especially water and land, over management of migration, or over compensation payments between countries mainly responsible for climate change and those countries most affected by its more destructive effects.

The report delineates four "conflict constellations" possibly emanating from global warming:

- Conflict constellation due to "climate-induced degradation of freshwater resources"

- Conflict constellation due to "climate-induced decline in food production"
- Conflict constellation due to "climate-induced increase in storm and flood disasters"
- Conflict constellation due to "environmentally induced migration"

The council identifies the following "regional hot spots" in terms of the one or more of the aforementioned potential conflict constellations: North Africa, the Sahel, southern Africa, Central Asia, South Asia (particularly India, Pakistan, and Bangladesh), the Caribbean and Gulf of Mexico basins, and the Andean/Amazonia region. They also delineate six key threats to international security and stability that could arise if mitigation of global warming fails:

- A possible increase in the number of weak and fragile states resulting from global warming
- Risks to global economic development
- Risks of growing "international distributional conflicts" between the countries contributing the most to global warming and countries most adversely impacted by it
- Risks to human rights and "industrialized countries' legitimacy as global governance actors"
- Impulse on the part of people in regions most adversely affected by global warming to migrate to regions less adversely affected by it

Dan Smith and Janani Vivekananda (2007:3) wrote a report in which they assert that many of world's poorest countries and communities face a "double-headed" dilemma: global warming and the potential for violent conflict. Like others, they maintain that global warming "could compound propensity for violent conflict, which in turn will leave communities poorer, less resilient and less able to cope with consequences of climate change" (Smith and Vivekananda 2007:46). Smith and Vivekananda submit that the 46 countries where the "effects of global warming interacting with economic, social and political problems could create a high risk of violent conflict" are home to some 2.7 billion people and that the 56 countries "where governments will have great difficulty in taking the strain of climate change on top of all their other current challenges" are home to some 1.2 billion people. Their report includes case studies of Algeria, Peru, Bangladesh, Mali and Chad, Liberia, and Nepal. The *Christian Science Monitor* has identified six potential flash points that could erupt into conflict as a result of global warming: Nepal, Indonesia, Lagos (Nigeria), the United States, the Arctic, and East Africa (Shapley 2007).

In 2008 the High Representative and the European Commission to the European Council (2008:1) released a report stating that "unmitigated climate change beyond 2°C will lead to unprecedented security scenarios as it is likely to trigger a number of tipping points that could lead to further accelerated, irreversible and largely unpredictable climate changes." This report anticipates the possibility that climate change "will fuel the politics of resentment between those most responsible and those most affected by it" and recognizes that droughts and food insecurity induced by climate change in sub-Saharan Africa may result in the intensification of efforts on the part of climate refugees to migrate to Europe (High Representative and the European Commission to the European Council 2008:5). It goes on to recommend that the UN Security Council, other UN bodies, and the G8 "enhance international cooperation and monitoring of the security threats related to climate change, and . . . prevention, preparedness, mitigation and response capacities" (High Representative and the European Commission to the European Council 2008:10).

More recently, the National Intelligence Council in the United States released a briefing document asserting that climate change will "exacerbate internal state pressures, and generate interstate friction through competition for resources or disagreements over responses and responsibility for migration" (quoted in Gilding 2011:109). Furthermore, in 2010 the Pentagon's Quadrennial Defense Review observed that climate change will constitute an "accelerant of instability or conflict, placing a burden to respond on civilian institutions and militaries around the world" (quoted in Gilding 2011:109).

Winners and Losers in the Era of Climate Change

Some of the discussion about the impact of climate change on human societies is framed in terms of neoclassical microeconomics or the neoliberal discourse of winners and losers. For example, Ward (2010:174) asserts,

> Among the winners will be locales that today are too cold to be desirable for year-round dwelling. Through geographic accident, most such places are in the Northern Hemisphere. The biggest victors will be Canada, Alaska, Greenland, Russia, and Scandinavia, and in the Southern Hemisphere, Argentina most of all. Perhaps future world power will not be relocated to the countries in the Southern Hemisphere, as is often predicted, but will stay concentrated, if not redistributed, in the Northern Hemisphere.

Some analyses even posit winners and losers within a specific region or locale. Remotely located Tasmania and New Zealand are also sometimes mentioned as winners. Conservative Canadian prime minister Stephen Harper periodically expresses excitement about visions of an economically prosperous northern frontier.

The executive summary of the 2004 *Arctic Climate Impact Assessment* maintains that the "reduction in sea ice is very likely to have devastating consequences for polar bears, ice-dependent seals, and local people for whom the animals are a primary food source," but it adds that "increased areas of tree growth for the Arctic could serve to take up carbon dioxide and supply more wood products and related employment" (Hassol 2004). Peter Jull (2009–2010:45) observes that dramatic climatic changes impacting the Arctic sea ice and Greenland glaciers

> have spawned an explosion of interest and activity relating to Arctic shipping, whether through the High Arctic islands (Lancaster Sound and the Northwest Passage) or even across the open Arctic Ocean near the North Pole. The Northeast Passage, around Norway and across the top of Russia to the Far East, is also in play, as the Russians have much more experience in Arctic shipping.

Both Democratic and Republican politicians from Alaska view climate change as a factor that potentially will spur on even further development, including oil and natural gas exploration in the North Slope (Emmerson 2010:256–57). One study estimates that the Arctic may contain as much as 13 percent of the world's oil deposits and 30 percent of its natural gas deposits (Howard 2009:8). Laurence Smith (2011:187) argues that a U.S Geological Survey in 2009 indicates that northern Alaska is the big winner in terms of oil and Russia in terms of natural gas in that the "Alaska Platform . . . is thought to hold between 15 and 45 billion barrels of oil with a best guess of about 28 billion," and "Russia's South Kara Sea alone is thought to hold between 200 and 1,400 trillion cubic feet of natural gas, with a best guess of 207 trillion." While oil can be relatively easily transported in tankers, the transport of natural gas requires laying extensive pipelines or building gas-to-liquid conversion facilities, both highly expensive operations. At any rate, Statoil, a Norwegian oil company, is looking to the Arctic as a site for new oil sources, given that its existing fields have been declining (Emmerson 2010:260). Indeed, the Norwegian government has prepared a 140-page document that "amounts to a plan for Arctic development from Norway's northern coastline up to near 85°North, further north than any permanent human settlement on Earth" (Emmerson 2010:261).

As a result of retreating ice in Greenland, various mining companies are hoping to gain easier access to various minerals, such as molybdenum, lead, zinc, diamonds, and even uranium, which presently cannot be mined due to legal prohibitions (Emmerson 2010:296). The Greenlandic Bureau of Minerals and Petroleum envisions seven mines opening over the next several years, which would result in some 15,000 jobs in a country with approximately 56,000 permanent residents at present. For Greenland fishermen, "a warming Arctic climate could bring benefits—fish are heading north" (Emmerson 2010:315–16).

Icelandic hydroelectric facilities are anticipating increased production, at least in the immediate future, due to the acceleration of the melt rate of the country's glaciers (Emmerson 2010:330–31). The increased availability of electricity appears to have prompted Becromal, an Italian company, to develop a new aluminum-foil production plant in Akureyri, Iceland's second-largest city. As Russia's oil and natural gas reserves are depleting, the country's oil and natural gas industries, including Gazprom, are looking to the Arctic for new supplies (Emmerson 2010:231–38). Russian oil and gas development in the Arctic promises to serve as a boon to the port of Murmansk (population 350,000), which, situated at latitude 68° north, is the largest city above the Arctic Circle. In contrast to the Arctic, Antarctica and its surrounding seas are protected from mineral extraction until 2048 under the provisions of the 1991 Protocol on Environmental Protection to the Antarctic Treaty. Anthony Bergen, an Australian policy analyst, reports, however, that China in particular is very interested in exploiting Antarctica's natural resources and notes that "as the world's fossil fuels diminish and technology for mining in polar regions improves, pressure will grow for Antarctic mineral development" (quoted in Chandler 2011:222).

While it is predicted that tourism will be severely impacted by climate change in many parts of the world—such as at ski resorts in both the European and Australian Alps and in much of the Mediterranean and the South Pacific—there are indications that some forms of tourism are on the rise in the Arctic region. Since the 1970s, there has been an almost fivefold increase in the number of tourists to visit Sakha in the Russian Arctic. "Every year around 1.5 million visitors are taken by cruise ship, bus and plane to places that were considered impenetrable wildernesses not long ago" (Howard 2009:31).

Social Justice Initiatives

Over the past decade or so, various social justice initiatives have emerged that, although not seeking to transcend global capitalism per se, seek to make it both more socially just and environmentally sustainable, including in terms of climate change. Tom Athanasiou and Paul Baer (2002) contend that mitigation will require a total drop to 60 to 80 percent below 1990 levels of greenhouse gas emissions and advocate the implementation of a "precautionary global emissions cap." They maintain that the developed countries need to assist the developing countries in engaging in a rapid process of "leapfrogging" that will allow them to adopt more environmentally sustainable technologies. While not advocating an end to global capitalism per se, Athanasiou and Baer (2002:113) propose substituting "privatization of the commons by establishing the institutions and politics of communal ownership" with respect to the atmosphere. They agree with the position of the International Forum on Globalization, which argues, "There is an appropriate place for private ownership and markets to play in the management, allocation, and delivery of certain common heritage resources, as for example land, within a framework of effective democratically accountable public regulation that guarantees fair pricing, equitable access, quality, and public stewardship" (quoted in Athanasiou and Baer 2002:142). Unfortunately, despite their commitment to poverty alleviation, environmental sustainability, and democracy, Athanasiou and Baer (2002:142) fail to make clear how these ideals will be achieved within the parameters of even a regulated capitalist world system.

The South-North Dialogue on Equity in the Greenhouse delineates six groups of countries representing different stages of development and three criteria in terms of climate change mitigation policies: (1) historical responsibility, (2) capability (as measured by per capita GDP and human development index), and (3) potential to mitigate (Jarman 2007:92). Roberts and Parks (2007:144) delineate four methods of differentiating among countries on the issue of "cutting the carbon cake": The first would be "grandfathering," which would simply require countries to reduce their emissions incrementally from a baseline year, such as 1990 as stipulated in the Kyoto Protocol. The second approach evaluates countries in terms of their carbon intensity, a concept that was introduced by the World Resources Institute and favored by the George W. Bush administration starting in 2002. This approach calls upon governments and corporate entities to make voluntary improvements in energy efficiency to order to reduce emissions. The third approach argues that developed countries bear much more responsibility in climate change mitigation efforts than developing countries because historically they have emitted, and they continue to emit, much higher levels of greenhouse gases than do developing societies. In a similar vein, the Greenhouse Development Rights approach establishes indices of countries' responsibility for greenhouse gas emissions (Jarman 2007:104; Baer et al. 2009).

A fourth approach is known as contraction and convergence and asserts that every human on Earth has equal rights to global atmosphere and a right to pollute on a per capita basis. This approach has been favored by India, China, and the Group of 77, which actually consists of about 133 nations. It has also been endorsed by France, Switzerland, and the European Union, despite the fact that developed countries will have to drastically reduce their emissions because most of them have already exceeded the requisite stabilization targets. The contraction-and-convergence approach was first proposed by the Global Commons Institute (Jarman 2007:98).

Lester R. Brown (2009:23–24), director of the US-based Earth Policy Institute, has devised *Plan B*, a scheme designed to save civilization, which has four components: "cutting net carbon dioxide emissions 80 percent by 2020, stabilizing population at 8 billion or lower, eradicating poverty, and restoring the earth's natural systems, including its soils, aquifers, forests, grasslands, and fisheries." He argues that *Plan B* is an integrated program in that

> we are not, for example, likely to stabilize population unless we can eradicate poverty. Conversely, we cannot restore the earth's natural systems without stabilizing population and climate, and we are not likely to stabilize climate unless we also stabilize population. Nor can we eradicate poverty without restoring the earth's natural resources. (Brown 2009:25)

Brown's plan combines elements of social justice and ecological modernization and essentially constitutes yet another green social democratic approach. He utilizes the drastic government-initiated structuring of the US economy during World War II as a model of how to restructure the global economy or global capitalism. In terms of stabilizing the climate, Brown (2009:79) calls for an energy revolution that would entail a shift from an "economy powered by oil, coal, and natural gas to one powered by wind, solar, and geothermal energy." He also calls for improved energy efficiency in various arenas, including lighting, appliances, buildings, transport, and the production, processing, and disposal of materials (Brown 2009:79–108). Brown (2009:143–67) also advocates drastically redesigning cities in terms of transport, water utilization, and even farming and upgrading squatter settlements. Unlike many neo-Malthusians, he does recognize that population will not stabilize or diminish until poverty, particularly among women, has been eradicated (Brown 2009:168–91). Last but not least, Brown (2009:192–238) calls for systemic efforts to protect and restore forests (which could serve to sequester carbon), conserve and restore soils, regenerate fisheries, and protect plant and animal diversity. He seeks to accomplish these goals and others within the parameters of capitalism as is indicated by the following remarks:

> The key to building a global economy that can sustain economic progress is the creation of an honest market, one that tells the ecological truth. To create an honest market, we need to restructure the tax system by reducing taxes on work and raising those on carbon emissions and other environmentally destructive activities, thus incorporating indirect costs into the market price. (Brown 2009:243)

While many elements of Brown's *Plan B* are commendable and could even serve as transitional stages toward creating a democratic eco-socialist world system or form integral parts of it, his scheme, for example, does not challenge the treadmill of production and consumption and the need to constantly grow, which are integral components of global capitalism. Furthermore, although he acknowledges the need to eradicate poverty, his scheme does not even suggest the need for a drastic redistribution of wealth and a shift to pronounced social equity. In this regard, *Plan B* resembles other campaigns to eradicate poverty or "make poverty history."

In his book *Kyoto2*, Oliver Tickell (2008:8–9) delineates a scheme that incorporates social justice considerations but that could be achieved by adopting market mechanisms. The objectives in his *Kyoto2* framework consist of the following propositions:

- The progressive limitation of greenhouse gas emissions year by year in order to achieve global climate neutrality by 2050 and long-term greenhouse gas stabilization at no more than 350 ppm CO_2e
- A shift to a low-carbon economy based upon renewable energy and energy efficiency and assisting "energy-poor" countries
- Improvement of the quality of life of the poorest people in the world
- Provision of funds from climate change adaptation, particularly for the poorest people and poorest countries
- The provision of financial incentives to developing countries to ensure protection of their carbon-rich ecosystems, such as forests, swamps, and peat lands
- The promotion of agricultural reforms that would enhance the role of soils as sinks and long-term reservoirs of CO_2
- The transfer by developed countries of finance, technology, and knowledge to developing countries in order to achieve the aforementioned objectives

Murphy (2008:111–25) has drawn up a Plan C that includes the following components:

- A drastic reduction in the consumption of fossil fuel energy and fossil fuel–derived products
- A shift from a growing economy to a contracting economy
- An emphasis on small communities
- The consumption of less food, dietary changes, reduction in meat consumption, purchase of local organic food, preservation and storage of food, and creation of gardens and/or henhouses
- A shift to energy-efficient cars and sharing rides
- The erection of smaller homes

For the most part, the suggestions delineated in Plan C might be more appropriate for most people in developed countries than the more affluent sectors of developing countries. His call for a shift from a growing economy to a contracting economy is incompatible with global capitalism and thus would require a transcendence of it, something that Murphy does not explicitly suggest.

Some social justice initiatives have adopted the notion of *climate debt* as one component in a larger litany of ecological debts. Climate debt includes the following dimensions:

- Acknowledgment of the impacts of excessive greenhouse gas emissions leading to extreme and frequent climate events, floods, droughts, inundations, storms, loss of arable land and biodiversity, disease, landlessness, migration, poverty, and so forth
- Reorganization of societies and economies in such a way that their greenhouse gas emissions are radically reduced
- Implementation of an emissions debt that recognizes the fact that rich countries have used up most of the atmosphere's capacity to absorb greenhouse gases, leaving no "atmospheric space" for the South to "grow" (Bullard 2010)

The Bolivian government has added two other items to the climate debt calculation: (1) the migration debt, which would be compensated by dropping restrictive migration practices, and (2) the debt to Mother Earth.

The Need to Think Outside the Box: Toward a Democratic Eco-Socialist World System

While the powers that be around the world are seeking to address climate change within the parameters of global capitalism, as Simms (2009:184) observes, "global warming probably means the death of capitalism as the dominant organising framework for the global economy." Thus, it is imperative to think outside the box and construct an alternative to global capitalism as the ultimate climate mitigation strategy, even though it will not be achieved any time soon, if indeed ever. As humanity enters an era of dangerous climate change accompanied by tumultuous environmental and societal consequences, it will have to consider alternatives that hopefully will circumvent dystopian scenarios on the order of those delineated in the previous section. Thus, in this section I propose the creation of a democratic eco-socialist world system as what sociologist Erik Olin Wright (2010) terms a real utopia.

Despite all the baggage associated with the term *socialism* and the desire of various progressive thinkers to substitute terms such as *radical democracy*, *economic democracy*, and *global democracy*, it is important for socialists to grapple with the ideals of socialism and the social experiments that have been labeled socialist, both at the national and local levels. As Samir Amin (2009:22) so aptly asserts,

> The expression counterculture is fraught with difficulty—because socialist culture is not there in front of our eyes. It is part of a future to be invented, a project of civilization, open to the creativity of the imagination.

In other words, socialism remains very much a vision, one with which various individuals and groups continue to grapple, often by seeking to frame it in new guises. Nevertheless, as Stilwell (1992:211) argues, "liberated from [the] Stalinist legacy, it now makes sense to start asking what a progressive socialism involves." Shiva (2008:46) utilizes the term *Earth democracy*, which incorporates solutions that "are coming from those who know how to live lightly, who have never had an oil addiction, who do not define the good life as 'shop until you drop,' but rather define it as looking after the living earth and their living community." Furthermore, her concept is "based on equal rights of all beings to ecological space, including atmospheric space" (Shiva 2008:47).

Anthropologists have long recognized that social systems, whether local, regional, or global, do not last forever. Global capitalism has been around for some 500 years, but I believe that it must be transcended if humanity and other forms of life are going to survive in some reasonable fashion. Thus, we need to consider an alternative world system based on social parity or justice and environmental sustainability. I propose this scenario as an alternative to an eco-fascist, or at least eco-authoritarian, vision that proposes that a small global elite juxtapose environmental sustainability and climate change mitigation with ongoing social inequality, authoritarian statism, and a smaller population. Li (2009:1058) provides the following thoughts on the possible nature of a future postcapitalist society:

> The collapse of capitalism and the establishment of a post-capitalist society will not automatically guarantee the solution of the climate change crisis and a successful transition to ecological sustainability. However, without the compulsive competitive demands imposed by the global capitalist market, humanity will be freed from the constant and intense pressure of ceaseless accumulation. Humans will be in a position to apply their collective rationality. Hopefully, people throughout the world will engage in a transparent, rational and democratic debate which is open not only to economic and political leaders and expert intellectuals, but also to the broad masses of workers and peasants. Through such a global collective debate, a democratic consensus could emerge that would decide on a path of global social transformation that would in turn lead to climate stabilization and ecological sustainability.

In the nineteenth century, various revolutionaries and reformers sought to develop alternatives to an increasingly globalizing capitalist world system. Efforts at the national level to create such an alternative started out with the Bolshevik revolution in Russia in 1917 and included subsequent revolutions in other countries, including China in 1949, Vietnam in 1954, Cuba in 1959, and Nicaragua in 1979. Unfortunately, as Wright (2010:106) observes, "these attempts at ruptural transformation, however, have never been able to sustain an extended process of democratic experimentalist institution-building" for a variety of complex reasons.

Scholars have spilled much ink and used up printer cartridges trying to determine whether these societies constituted examples of "state socialism," "actually existing socialism," transitions between capitalism and socialism that required further democratization, "state capitalism," or "new class societies" and why many of these societies eventually became fully incorporated into the capitalist world system, beginning with the collapse of the Soviet Union and the Eastern bloc countries in the early 1990s. Suffice it to say that their failure to achieve authentically democratic socialist societies was ultimately related to both internal forces specific to each of these societies and external forces that created a hostile environment for equitable development. According to Wright (2010:106),

> Perhaps the failure of sustained democratic experimentalism in the aftermath of revolutions was because revolutionary regimes always faced extreme pressure, both economic and military, from powerful capitalist countries, and felt a great urgency to consolidate power and build institutions of sufficient strength to withstand that pressure. . . . Or perhaps the problem was mainly the low level of economic development of the economies within which revolutionary movements succeeded in seizing political power. Classical Marxism certainly never imagined that a transformation of capitalism into a democratic egalitarian alternative would be possible unless capitalism had already generated very high levels of productivity.

In a similar vein, Wallerstein (1998:12) asserts that revolutionary regimes "are constrained by the structures of the world-system to behave in certain ways and within certain parameters or else they lose all capacity to be important actors in the world-system."

The collapse of Communist regimes created a crisis for people on the left throughout the world. Many progressive people had hoped that somehow these societies would undergo changes that would transform them into democratic and ecologically sustainable socialist societies. Various progressives have advocated shedding the concept of socialism and replacing it with terms such as *radical democracy, economic democracy*, and *anticapitalist society* (Aronowitz 1994). While efforts to replace the term *socialism* with new ones are understandable given the fate of postrevolutionary or socialist-oriented societies, progressive people need to come to terms with both the achievements and flaws of these societies and to reconceptualize the concept of socialism. According to Ralph Miliband (1994:51), three core propositions define socialism: (1) democracy, (2) egalitarianism, and (3) socialization or public ownership of a predominant part of the means of production. Although some areas of a socialist society and ultimately world system would require centralized planning, coordination, and governance, democratic socialism recognizes the need for widespread decentralized economic, political, and social structures that would permit the greatest amount of popular participation in decision making possible. Wallis (2006:42) observes that the socialist banner includes an array of specific agendas. Socialist democracy would not be synonymous with total state ownership and centralized planning but could entail "several forms of property—collective, cooperative and small private or individual property" and even space for small businesses (Lorimer 1997:22).

Over the past two decades or so, leftists have become more sensitive to the environmental travesties that have occurred not only in both developed and developing capitalist societies but also in postrevolutionary societies. As a result of this, various leftists have sought to develop an eco-socialism (Resistance 1999; Foster 2000, 2009). Joel Kovel (2008:8) provides a compelling perspective on eco-socialism, observing,

> Where there was no inherent impulsion within first-epoch socialism to look beyond human welfare, ecosocialism entails a radical shift away from the anthropocentric attitude that holds humanity over nature. Plainly, if life is under threat by capital, then the threat applies to all lives. . . . The option for an ecocentric perspective entails a decentering from our narrow species interest toward a more universal perspective that encompasses the ecosphere: the plenum of ecosystems, and all creatures, which constitutes and frames human existence.

Merrill Singer, Ida Susser, and I, in our work in critical medical anthropology, have utilized the notion of *democratic eco-socialism,* which entails the following principles (Baer, Singer, and Susser 2003:356–59):

- An economy oriented to meeting basic social needs—namely, adequate food, clothing, shelter, and health care
- A high degree of social equality
- Public ownership of means of production
- Representative and participatory democracy
- Environmental sustainability

Democratic eco-socialism rejects a statist, growth-centered, or productivist ethic and recognizes that humans live on an ecologically fragile planet with limited resources that must be sustained and renewed as much as possible for future generations.

Our vision of democratic eco-socialism resembles what world systems theorists Terry Boswell and Christopher Chase-Dunn (2000) term *global democracy,* which would entail the following components: (1) an increasing movement toward public ownership of productive forces at the local, regional, national, and international levels; (2) the development of an economy oriented toward meeting social needs, such as basic food, clothing, shelter, and health care, and environmental sustainability rather than profit making; (3) a blending of both representative and participatory democratic processes; (4) the eradication of health and social disparities and the redistribution of human resources between developed and developing societies and within societies in general; (5) the curtailment of population growth that in large part would follow from the previously mentioned condition; (6) the conservation of finite resources and the development of renewable energy resources, such as wind, solar, and geothermal energy; (7) the redesign of settlement and transportation systems to reduce energy demands and greenhouse gas emissions; and (8) the reduction of waste through recycling and transcending the reigning culture of consumption.

Many have argued that socialism has been tried in places like the Soviet Union and China, and even Cuba for that matter, and proven wanting. Of these three societies, Cuba is the closest example of an existing society that embodies socialist ideals and practices; yet, socialism, not to mention democratic eco-socialism, remains a vision rather than an existing social system per se. Nevertheless, developments in Latin America, particularly Venezuela, Bolivia, and certainly Cuba, raise the hope of creating a "socialism for the 21st century" (Katz 2007). As John Bellamy Foster (2009:276) so aptly argues,

> It is important to recognize that there is now an *ecology* as well as a political economy of revolutionary change. The emergence in our time of sustainable human development, in various revolutionary interstices within the global periphery, could mark the beginning of a universal revolt against both world alienation and human self-estrangement. Such a revolt, if consistent, could have only one objective: the creation of a society of associated producers rationally regulating their metabolic relation to nature, and doing so not only in accordance with their own needs but also those of future generations and life as a whole. Today, the transition to socialism and the transition to an ecological society are one.

While, at the present time or for the foreseeable future, the notion that democratic eco-socialism may be eventually implemented in any society, developed or developing, or in a number of societies may seem utterly ridiculous, history tells us that social changes can occur very quickly once economic, political, and social structural changes have reached a tipping point.

Conclusion

The vision of democratic eco-socialism provides people everywhere with an alternative to the existing capitalist world system that continues to self-destruct because of its socially unjust and environmentally unsustainable commitments and practices. Ultimately, the shift to democratic eco-socialism in any country would have to be part of a global process or a "permanent revolution" that no one fully envisions. The history of the Soviet Union and Stalinism tells us that socialism cannot be created in "one country." The struggle for a safe climate needs to be part and parcel of a larger struggle for social justice and environmental sustainability, both internationally and within specific nation-states. As Magdoff and Foster (2010:25) so aptly argue, "Everywhere radical, essentially anti-capitalist, strategies are emerging, based on other ethics and forms of organization, rather than the profit motive: ecovillages; the new urban environment promoted in Curitiba in Brazil and elsewhere; experiments in permaculture, and community-supported agriculture, farming and industrial cooperatives in Venezuela, etc."

Samir Amin (2008:78–81) has proposed the creation of a Fifth International, which could draw on the conference at Bamako organized on January 18, 2006, on the eve of the opening of the 2006 Polycentric World Social Forum. The Bamako Appeal is based upon the following principles (cited in Amin 2008:108–11):

- "Construct a world based on solidarity among human beings and peoples."
- "Construct a world based on the full and complete citizenship and equality between the sexes."
- "Construct a universal civilization that offers the greatest possibility for the creative development of diversity in all areas."
- "Construct socialization through democracy."
- "Construct a world based on the recognition of the non-commodity status of nature, the planet's resources, and agricultural lands."
- "Promote policies that closely combine unlimited democracy, social progress, and the affirmation of the autonomy of nations and peoples."
- "Affirm the solidarity of the peoples of the North and South in the construction of internationalism on an anti-imperialist foundation."

By and large, the principles embodied in the Bamako Appeal closely parallel those found in democratic eco-socialism or global democracy. More recently, Fred Magdoff (2011:20) succinctly has delineated the principles guiding the formation of an *ecological civilization*: "It must (1) provide a decent human existence for everyone: food, clean water, sanitation, health care, housing, clothing, education, and cultural and recreational possibilities; (2) eliminate the domination or control of humans by others; (3) develop worker and community control of factories, farms, and other workplaces; (4) promote easy recall of elected personnel; and (5) re-create the unity between humans and natural systems in all aspects of life, including agriculture, industry, transportation, and living conditions." In terms of achieving such an ecological society, it would "stop growing when basic needs are satisfied; (2) not entice people to consume more and more; (3) protect natural life-support systems and respect the limits to natural resources, taking into account needs of future generations; (4) make decisions based on long-term societal/ecological needs, while not neglecting short-term needs of people; (5) run as much as possible on current (including recent past) energy instead of fossil fuels; (6) foster the human characteristics and a culture of cooperation, sharing, reciprocity, and responsibility of neighbours and community; (7) make possible the full development of human potential; and (8) promote truly democratic political and economic decision making for local, regional, and multiregional needs" (Magdoff 2011:20). Inspired in part by the principles delineated by Magdoff, in the next chapter I explore some of the specifics of achieving an ecological civilization or global democratic eco-socialism.

7

Toward an Ecological Revolution: Progressive Transitional Reforms

Obviously the transition toward a democratic eco-socialist world system is not guaranteed and will require a tedious, even convoluted path. Nevertheless, in the process of struggling for major social transformation, progressives can work on various transitional, or what Andre Gorz (1973) terms *nonreformist*, reforms. David Schwartzman (2009:19) advocates a solar-powered civilization, which, he argues, "will require using fossil fuel, the dominant energy source now available, for the creation of an alternative infrastructure." At the same time, carbon emissions will have to be minimized during the transition to a solar civilization. Schwartzman (2009:27) delineates the following as essential features of an eco-socialist transition:

1. Application of the containment and precautionary principles to environmental policy (including industrial ecology and organic agriculture centered around and in green cities);
2. Progressive dematerialization of technology and global availability of state-of-the-art information technology; and
3. Increase of human population density centered in green cities with elimination of sprawl, leaving extensive biospheric reserves that are managed to preserve biodiversity.

In large part, efforts to shift the global political economy from a capitalist framework to a democratic eco-socialist one will require a broad-based alliance of antisystemic movements, including those drawn from the labor movement, the global justice or anti–corporate globalization movement, the recent Occupy movement, radical political groups, indigenous and ethnic rights movements, the women's movement, and the environmental movement, including a relatively recent climate movement, which I discuss in the next chapter. Ultimately, such an alliance would have to attain state power to implement progressive policies and practices essential to the creation of an alternative world system based on social justice and environmental sustainability. Furthermore, as Foster (2009a:276) so aptly observes, "It follows that there is little real prospect for the needed global ecological revolution, unless these attempts to revolutionize social relations in the struggle for a just and sustainable society, now emerging in the periphery, are somehow mirrored in movements for ecological and social revolution in the advanced capitalist world."

In this chapter, I delineate several progressive transitional reforms essential to implementing an ecological revolution and ultimately democratic eco-socialism. These include the following: (1) the creation of new left parties; (2) the implementation of emissions taxes; (3) the nationalization or public ownership of the means of production; (4) increasing social equality; (5) the implementation of workers' democracy; (6) the shortening of the workweek; (7) the implementation of renewable energy sources, energy efficiency, and appropriate technology and the creation of green jobs; (8) the expansion of public transport; (9) the creation of green cities; (10) resistance to the culture of consumption; and (11) the implementation of sustainable agriculture and forestry. This list by no means exhausts the steps necessary of bringing about a socially just and environmentally sustainable world system. The details involved are much more complex, but hopefully my proposed transitional reforms will prompt a dialogue as to the possibilities involved in creating a "real utopia" that would be part and parcel of achieving a safe climate. While I suggest a litany of possible transitional reforms, I am not calling for a definitive blueprint for achieving an ecological revolution. The application of my suggested transitional reforms will have to be adapted for the many countries, both developed and developing, around the world. Furthermore, my suggested transitional reforms do not exhaust the litany of possible transitional reforms necessary for creating an alternative world system.

New Left Parties

Unfortunately, around much of the world today, to a greater or lesser degree, multinational or transnational corporations tend to make or break governments and politicians. Noam Chomsky has asserted the United States has a "one party system—the business party," with two factions, the Republicans and Democrats. In the case of Australia, the two major parties—the Coalition (Liberal and National parties) and the Australian Labor Party (ALP)—have also become factions of the "business party." Fortunately, a system of proportional representation and preference voting makes it more possible for minor parties, such as in the past the Democrats and now the Greens, to win seats in the Senate. Indeed, in the 2010 federal election, Adam Bandt, a Green, won the seat of Melbourne in the lower house, or the House of Representatives. The "winner-take-all" system in the United States makes it difficult for an independent or minor-party candidate to be elected to either the House of Representatives or the Senate. A notable exception is Bernie Sanders, an independent socialist from the small New England state of Vermont who served as mayor of Burlington, then several terms as a congressperson, and now is a US senator. In contrast a system of proportional representation makes it possible for smaller parties to win representation in legislatures or parliaments. For example, the German parliament, or Bundestag, has representatives from six political parties. Running from the far right to the far left, they are the Republicans, the Christian Democrats, the Free Democrats, the Social Democrats, the Greens, and Die Linke, or the Left Party. Since 2005, the Left Party has been represented in the Bundestag and started out as a coalition of Social Democratic dissidents, left trade unionists, and the Party of Democratic Socialism, or the successor party of the former Socialist Unity Party, the ruling party of the German Democratic Republic (Solty 2008:4).

Rosa Luxemburg viewed "parliamentary elections as an opportunity for the powerful development of socialist propaganda and for the assessment of socialist influence among the masses" (Froelich 1972:62). Social movement activists have been faced with the question once again of what prospects there are for the emergence of new political institutions that will carry an anticapitalist political agenda into twenty-first century (Panitch 2008:5). Socialist governments in various countries ultimately need to be part and parcel of a world government, with reduction of armies, police, and prisons (Sherman 1995:335).

Emissions Taxes

I propose an emissions tax, or more specifically a carbon tax, with great reluctance because of the tendency of the corporate class to find loopholes and pass the tax burden onto working-class people. An increasing number of scholars, policy advisors, and climate activists are advocating an emissions, or more specifically a carbon, tax. James Hansen and others have proposed that a "carbon tax . . . be implemented at the wellhead, mine gate or port of entry," given the flaws in other market mechanisms, particularly emissions trading schemes and carbon offsetting (Frank 2009:36). In 1993 the Bill Clinton administration proposed a British thermal unit–based tax on energy fuels. The proposal met with fierce opposition and was defeated in the Senate Finance Committee (Brohé, Eyre, and Howarth 2009:155). Yale economist William Nordhaus and climate scientist James Hansen maintained at the climate change conference in Copenhagen in March 2009 that carbon taxation was a crucial measure to be employed in addressing climate change (Brohé, Eyre, and Howarth 2009:292). Sociologist Anthony Giddens (2009:145) argues that "carbon taxes should be used to transfer the tax burden away from labour and toward taxing the sources of environmental pollution." He goes on to maintain that "taxes on the use of resources should be as near to the point of production as possible, in order to apply to all relevant aspects of manufacturing processes" (Giddens 2009:150).

Emissions taxes have the following advantages:

- They provide a clear cost penalty to reduce emissions and can be set at a steep rate and increased over time.
- They are relatively easy to administer and can be readily incorporated into existing corporate tax structures, which are too low given all the benefits that corporations obtain from government.
- They can provide governments with a significant source of revenue that can be funneled into other funds, such as developing renewable sources of energy and providing rebates for low-income people who will be adversely impacted by rising energy costs.

Conversely, emissions taxes have the following disadvantages:

- They do not guarantee a specific emission reduction outcome and thus must be coupled with other ways of reducing emissions, though certainly not an emissions trading scheme as these have repeatedly proven problematic.
- They have faced, and can be expected to face, opposition from corporate interests and their political allies.
- They are not a panacea but at best a flawed transitional reform that can be phased out with full public ownership of the means of production.
- They have the potential, like most tax schemes, to affect most adversely the poor or low-income groups.

Advocacy of emissions taxes or carbon taxes is very unpopular among conservative forces. Obviously, emissions tax schemes would be very difficult to implement. According to Brohé, Eyre, and Howarth (2009:293), there is "currently little experience with an international regime of taxation, and countries already have vastly different levels of fuel taxation that would be difficult to reconcile in practice under such an international system." However, in reality, climate-related regulation and tax codes already exist in various countries (Lohmann 2006:334). Denmark, Sweden, Norway, Finland, and Iceland introduced taxes on electricity, energy consumption, and fossil fuels beginning in the early 1990s. Emissions taxes are not a panacea in terms of climate change mitigation. Efforts to implement carbon taxes in Sweden, Finland, the Netherlands, Denmark, Germany, Norway, Italy, and some local areas in the United States have had mixed results. Chivers (2009:204) argues that the "taxes do seem to reduce carbon emissions, but usually on a smaller scale than was hoped for—often due to loopholes and concessions by industry or angry consumers groups." According to Giddens (1007:151–52),

> In Finland, without the CO_2 tax, emissions would probably have been 2–3 per cent higher by the year 2000 than they turned out to be; in Sweden, Norway and Iceland the figure was 3–4 per cent. The absolute level of emissions, however, increased across the 1990s in all these countries. Only in Denmark did the absolute volume of CO_2 emissions fall. The reason is that the Danes directed the tax revenue to environmental needs—it was used to subsidize energy-saving practices.

Unfortunately, efforts to impose a carbon tax in the European Union met with failure because certain member states, such as the United Kingdom, capitulated to the claims of various industries that they would not be able to compete with non-EU industries (O'Riordan and Jordan 1999:84). In a society committed to moving toward social equity and environmental sustainability, hopefully such barriers would be transcended.

As noted, the poor both in developed countries and particularly in developing countries potentially could be the ones most adversely impacted by emissions taxes. Thus, it is imperative that mechanisms be created that would minimize or, better yet, eliminate this, such as in the form of rebates (see Resistance 1999:157). For example, British Columbia's carbon tax scheme includes a rebate for low-income people (Chivers 2009:204). A carbon tax would need to be imposed at the site of production and be relatively steep. One way to prevent corporations from passing the costs of a carbon tax on to consumers would be stringent price regulations, such as in the case of the cost of electricity generated by a coal-fired or natural gas power plant.

Richard Cooper (2007:112) proposes a global carbon tax of $50 per ton of carbon, which "would amount to nearly $14 per ton of CO_2," and reviewing the carbon tax periodically to take into "account both greater knowledge about the impact of the tax and about the evolution of climate in response to continuing [greenhouse gas] emissions." Besides the advantages and disadvantages of a carbon tax, what is drastically needed is the abolishment of fossil fuel subsidies. Such subsidies currently come to US$80 billion in Organization for Economic Cooperation and Development (OECD) countries and US$20 billion in non-OECD countries (Barbier 2010:67).

Public Ownership of the Means of Production

In an era of increasing privatization of social and health services around the world, raising the specter of public ownership, or socialization of the means of production, is anathema in conventional economic and political circles. Privatization is often justified in terms of efficiency, the conventional argument being that private enterprises are more efficient and consumer-friendly than public or government ones. While government enterprises or services can be terribly inefficient for complex reasons, such as corruption and worker alienation, this does not have to be the case. Public ownership is not a particularly radical idea, not that I am opposed to radical ideas. Australia, for example, historically exhibited extensive public ownership of various productive forces, not only utilities but also banks, manufacturing operations, communication networks, Qantas Airlines, and transport systems. Robert Menzies, a conservative prime minister, reportedly stated in 1943, "Few people would have any quarrel with government control of railways, or tramways, or water supply, or such other great public utilities" (quoted in Wettenhall 1965:428). Socialization of private wealth would constitute an essential step toward the creation of a genuine democratic society and would reduce the power of the corporate class and wealthy individuals to influence elections around the world through the support of selected candidates via campaign contributions, favorable media coverage, and even bribery. Wall (2010a:57) maintains that "ecosocialism is founded on the principle of common property rights, which allow individuals [and groups] free access to a resource as long as they don't damage it."

Howard Sherman (1995:332–33) proposes a mix of public, cooperative, and public ownership as a means of facilitating economic democracy in the United States under which the biggest 100 firms would be publicly owned, the next 900 biggest firms would be employee owned, and the remainder would be privately owned. In a similar vein, Wall (2010a:60) observes, "Cooperatives exist in many parts of the world and provide democratic ways of producing and distributing goods and services. New commons-based forms of peer-to-peer production are also growing."

Socialization or public ownership, even in the form of cooperatives, has the potential of facilitating the shift from fossil fuel energy sources to renewable energy sources and what has come to be known as a "just transition" for workers who will be displaced in the former and transitioned to the latter. Capitalist owners of renewable energy companies have the potential to be just as exploitative as those in the fossil fuel industries historically have been. As Sergio Oceransky (2010:505) observes,

> Conflicts around the control of land, water, forests, and other ecosystems are seldom analysed in connection with the transition to renewable energy. As a consequence, there has not been much discussion about the territorial, economic, and cultural conflicts that are likely to be associated with a shift from fossil fuels to "new" renewable energies (wind, solar, wave, tidal, etc.) as the driving force behind industrial development.

In the case of Denmark, initially the national and local governments supported cooperatively owned windmills, which came to include more than 150,000 families. Unfortunately, due to governmental neoliberal practices, ownership later shifted away from cooperatives to private companies and investors (Kruse 2010:523). Ultimately, as Paul D'Amato (2006:138) observes, "nationalization can only be a weapon in the transformation of society in a socialist direction if the working class has first placed itself in power." He goes on to argue that the "aim of workers' power is to implement a series of economic and social transformations that do away with all class distinctions" (D'Amato 2006:138).

Workers' Democracy

A workers' democracy would constitute an integral part of a shift to democratic eco-socialism. Enterprise managers and directors could "be chosen (and periodically replaced) by lot, so that they are proportionately representative of the various grades of workers within the organisation, or directly elected by, and answerable to, the whole work force" (Mann 2010:259). Democratic planning needs to be part and parcel of the production process, such as in deciding what goods are needed and whether they are environmentally sustainable. As Lowy (2006:299–300) observes,

> Socialist planning must be grounded on a democratic and pluralist debate, at all the levels where decisions are to be taken. As organized in the form of parties, platforms, or any other political movements, delegates to the planning bodies are elected, and different propositions are submitted to all the people concerned with them. . . . Should public transportation be free? Should the owners of private cars pay special taxes to subsidize public transportation? Should solar energy be subsidized, in order to compete with fossil energy? Should the work week be reduced to 30 or 25 hours, or less, even if this means a reduction of production? The democratic nature of planning is not incompatible with the existence of experts: their role is

> not to decide, but to present their views—often different, if not opposite—to the democratic process of decision-making.

In a similar vein, Maurice Brinton (1975), a member of the London libertarian socialist group Solidarity, argues that the working class must gain direct power of production activities in order to have power in any society. Michael Albert and Robin Hahnel (1991) have developed an elaborate model of participatory economics, which they call "parecon," that would entail a network of workplace and consumer-based councils. It is a libertarian socialist model that "entails participatory planning, which is a kind of cooperative, horizontal negotiation of inputs and outputs, balanced job complexes in which each worker gets a combination of responsibilities so that his or her overall workload is empowering compared to what others enjoy, and remuneration for how long people work, how hard people work, and the onerousness of the conditions under which people work" (Albert 2008:143–44). While parecon has been criticized from various quarters for not being sufficiently sensitive to environmental factors, Robin Hahnel (2008:72) recently observed that participatory economics will need input from environmentalists:

> An active environmental movement will be necessary in a participatory economy to argue for the importance of environmental protection and restoration. . . . Environmentalists will have to speak up in worker and consumer councils and federations, pointing out the true benefits of environmental preservation and the magnitude of environmental degradation.

Increasing Social Equality

The United Nations Development Programme (UNDP) has consistently recognized widespread social inequality in the world. For example, in its 2007–2008 report, it states,

> The 40 percent of the world's population living on less than US$2 a day accounts for 5 percent of global income. The richest 20 percent accounts for three-quarters of the world income.

> In the case of sub-Saharan Africa, a whole region has been left behind: it will account for almost one-third of world poverty in 2015, up from one-fifth in 1990. (UNDP 2007:25)

UNDP (2007:25) also recognizes that "more than 80 percent of the world's population lives in countries where income differentials are widening." While organizations such UNDP and the World Bank claim to be committed to the eradication of global poverty, they rarely, if ever, call for a redistribution of wealth or increasing social equality. Instead such organizations call for more economic growth so that the poor will be delivered from their poverty, never recognizing that wealth and poverty, development and underdevelopment, are intricately interwoven phenomena. While some redistribution of wealth has at times been achieved under capitalism, more so within particular nation-states than between nation-states, social inequality is an inevitable component of the capitalist world system.

Ultimately, a shift toward greater social equality or parity will require transcending global capitalism and moving toward a democratic eco-socialist world system. In contrast to communism, which is based on the adage "From each according to his abilities, to each according to his needs," socialism, as a transitional phase between capitalism and communism, is based on the adage "From each according to his abilities, to each according to his work." Thus, under socialism, individuals who work longer, have greater abilities and skills, and take on more responsibilities are rewarded more than others. Socialists have over the years engaged in intense debates about what sort of wage differentials should exist in a socialist society. Stilwell (2000:130) argues that a 3:1 ratio of the highest to lowest incomes would be a tolerable standard for a socialist society. Other socialists, however, argue for a somewhat larger income differential on the grounds that some individuals may choose to work many more hours than others to obtain certain expensive consumer items, and others may choose to work a very short period so as to pursue various avocations. Conversely, it is important to note that there are other compensations for meaningful work than material rewards, such as the intrinsic rewards of intellectual and even physical stimulation and the sense that one has contributed to the greater good. Besides the ideal wage differentials under socialism, there is no question that they should be significantly less than they are in all capitalist societies, including Japan and the social democracies of Scandinavia. Özlem Onaran (2010:19) argues that in order to be socially acceptable, maintaining "long-term economic growth at zero or low levels" will require a "guarantee of high employment and an equitable distribution of income," all of which are incompatible with capitalism.

Various measures have been designed to counter the shortcomings of gross national product or gross domestic product in assessing the socioeconomic prosperity or well-being of countries. Following none other than Marx himself, Michael Lebowitz (2010:13) argues that a "good society is one that permits the full development of human potential." The Human Development Index (HDI) is designed to incorporate life expectancy, knowledge, standard of living, and other indicators. The Gross National Happiness Index was created by King Jigme Singye Wangchuk in 1972 (Sim 2010a:155). A University of Leicester study conducted in 2007 indicated that Bhutan is eighth out of 178 countries in terms of subjective well-being and outscored the United States, the United Kingdom, and Canada. Denmark and Switzerland tied for first (Sim 2010b:156). Unfortunately, HDI and some other measures fail to consider environmental sustainability.

The Yale Center for Environmental Law and Policy and the Center for International Earth Science Information Network at Columbia University developed an Environmental Sustainability Index that "explicitly combines social and environmental variables grouped in five components: (1) The existing quality of the environment, including biodiversity, land, and water; (2) Efforts to reduce environmental stresses, such as lowering population growth, pollution, and the ecological footprint; (3) Efforts to reduce human vulnerability, by improving nutrition and public health, and reducing exposure to environmental hazards; (4) Social and institutional capacity to set and enforce environmental standards, and the technical basis to manage resources and improve efficiency; and (5) Global stewardship, especially international cooperation to reduce global warming and manage cross-national environmental issues"(Bodley 2008:306–7). The New Economic Foundation has a Happy Planet Index, incorporating ecological footprint, life satisfaction measures, and life expectancy, that seeks to determine how efficiently countries are utilizing natural resources to produce "happy life years" (Sim 2010b:158). Costa Rica has the highest index with 99 percent renewable energy, life expectancy of 78.5 years, and an average satisfaction score of 8.5 out of 10. Bhutan scores 13 out of 178 countries. Ironically, the United States, supposedly the richest country in the world, comes in at 114, "largely because its ecological footprint is so high relative to 'happy life years' results that are about average for wealthy countries" (Schor 2010:179).

Many mainstream environmental activists, including ones in the climate movement, often posit population growth as a major culprit contributing to environmental degradation, including climate change. In reality, creating a more even playing field in terms of access to basic resources would eventually eradicate the population bugaboo. Despite persistent alarm about population growth, the annual rate of growth has risen and fallen over the course of the past 60 years. It increased from about 1.5 percent in 1950 and 1951 to more than 2 percent in the early 1960s, stabilized around 2 percent in the early 1970s, dropped to 1.4 percent several years ago, and presently stands at 1.2 percent (Poston and Bouvier 2010:265). Ultimately, as Bodley (2008:218) asserts, "a more equitable distribution of economic resources would help slow population growth; however, it is crucial that improvements in the economic conditions of households be accompanied by policy changes that give women more control over fertility decision making."

A Shorter Workweek

Despite the historical fact that Australia invented the 40-hour week, most full-time employed Australians in various occupations and professions are working over 40 hours per week, often well over. In a similar vein to the United States, which has become the land of what radical Harvard economist Juliet Schor (1992) calls the "overworked American," the land down under has become a country of "overworked Australians." Whereas full-time workers worked 44 hours per week in both Australia and New Zealand in 2007, those in the United Kingdom worked 43 hours; in France and Germany, 41 hours; in Denmark, Finland, and Sweden, 40 hours; and in the Netherlands and Norway, 39 hours (Tiffen and Gittins 2009:82). Comparing average weeks worked in 2005, Australian workers put in 46.0 weeks per year, only slightly less than American workers, who put in 46.2 weeks, but more than workers in most other developed countries. In 2005, Irish workers put in 43.9 weeks; Italian workers, 41.3 weeks; German workers, 40.6 weeks; Norwegian workers, 37.0 weeks; and Swedish workers, 36.0 weeks (Tiffen and Gittins 2009:82). The reason that workers in various Scandinavian countries work fewer hours per week and fewer weeks per year than those in Australia appears to be related to the fact that the former have a higher percentage of workers belonging to trade unions. Whereas in 2003 23 percent of Australian workers belonged to trade unions, 78 percent of those in Sweden, 74 percent of those in Finland, and 53 percent of those in Norway belonged to trade unions (Tiffen and Gittins 2009:86). As part of resisting the culture of consumption, a return to a more laid-back Australia with reduced working hours may prove an important climate mitigation strategy.

Renewable Energy Sources, Appropriate Technology, and Green Jobs

Obviously, a shift to renewable energy sources and energy efficiency constitutes an important component of climate change mitigation, and various examples already exist that illustrate that such as shift is viable. For example, with support from the Ministry of the Environment, as of 2008, nearly 14 percent of Germany's electricity came from renewable energy sources

(Rogers 2010:86). Germany plans to obtain 25 to 30 percent of its energy from renewable sources by 2020.

Renewable sources of energy are a vital component of climate change mitigation but not a panacea, as Ross McCluney (2005:172) aptly observes:

> Despite the attractive energy potentials of the world's wind, waves, ocean currents, tidal currents, and geothermal sources, extracting significant portions of this energy presents environmental, economic, and social problems. Fortunately most are relatively easy to overcome when the power plants are small or widely separated. Modest use of these renewable technologies may be easily tolerated. However, considering the growth in world population and rising expectations for plentiful energy, future demand will put great pressure on developers to capture as much of each resource as possible, perhaps with terrible environmental consequences.

Ted Trainer (2007:2), a social scientist and research fellow at the University of New South Wales, acknowledges the superiority of renewable sources of energy over fossil fuels but maintains that the "very high levels of production and consumption and therefore of energy use that we have in today's consumer-capitalist society cannot be sustained by renewable sources of energy." In the nineteenth century, William Stanley Jevons observed that an increase in the energy efficiency of the steam engine led to an increase in coal consumption. According to what has come to be known as the Jevons paradox, conserving energy by improving efficiency within the context of a capitalist system actually increases its use. Trainer's reservations about renewable sources of energy make him especially concerned to find ways to restrain capitalism's tendency toward overproduction and the associated encouragement of consumerism, which has been so destructive of the planetary environment. Trainer argues that the green movement in general is deeply flawed and is "for the most part only light green." He argues that the green movement tends to make faulty assumptions about the degree to which renewable sources of energy can contribute to an environmentally sustainable world. Trainer asserts that wind energy has the potential to make an appreciable contribution to the electricity supply in many regions and countries, including in Europe, Canada, the United States, New Zealand, and portions of Australia; however, considering the world as a whole, wind energy can meet only a small portion of the demand for electricity (Trainer 2007:40, 41). In his view, solar thermal systems can provide electricity during summers in the hottest regions with the use of very long transmission lines, but such technology is not very effective during the winter, particularly in temperate regions (Trainer 2007:57). Photovoltaic solar energy, he believes, can play a part in the development of a renewable energy economy, but it poses technical problems in providing electricity at night, in cloudy weather, and during winter (Trainer 2007:72). In general, he insists that renewable energy cannot contribute significantly to providing current levels of transport (Trainer 2007:91). He argues that tidal energy at this time would prove very costly, and while relying on ocean currents to propel turbines sitting in locations with strong currents appears promising, not much energy could be obtained in this manner (Trainer 2007:107). Only time will tell whether geothermal sources of energy will prove practical and relatively inexpensive (Trainer 2007:109).

While both renewable energy sources and improving energy efficiency can play an important role in climate change mitigation, they are obviously not panaceas if not coupled with a commitment to breaking the treadmill of production and consumption and ongoing economic expansion. Trainer calls for "appropriate development" for both "rich" and "poor countries." For poor or developing countries, appropriate development "focuses on developing what is needed, and that is totally different from development defined as facilitating whatever will maximise GDP or business turnover" (Trainer 2011:130). Appropriate development entails local economic self-sufficiency, the utilization of alternative technologies reliant as much as possible on local resources, and environmental sustainability. For rich or developed countries, appropriate development would entail an enormous reduction in consumerism (Trainer 1989:196), ideally a "zero-growth economy" in which societies "will work hard at reducing the amount of producing and consuming going on" (Trainer 1995:108). In a somewhat similar vein, Altavater (2006:54) maintains that "the transition to renewable energy requires appropriate technologies, but requires even more appropriate social institutions and economic forms."

A shift from the present capitalist world system will have to entail a shift to green jobs, ones that are not only environmentally sustainable but also cater to people's social and health-care needs. Onaran (2010:30) calls for "public expenditures in labor-intensive services like education, child care, nursing homes, health, and community and social services." The creation of green jobs must be accompanied by what has become called a "just transition." Conversely, as Brian Kohler (2010:572), a spokesperson for the International Federation of Chemical, Energy, Mine and General Workers' Unions, cautions, "It would be very much easier to sell sustainability to trade unionists, especially trade unionists in dirty, toxic, or resource-depleting industries, if there were excellent examples of Just Transition to point to." He goes on to argue that governments or government-directed programs will be needed to facilitate this process.

Employment in the renewable energy sectors has been gradually increasing in various countries around the world, including China.

Biofuels in general tend to be problematic as a renewable source of energy because, as noted earlier, they require huge parcels of land that could be used for food production instead. Conversely, perhaps in time technological innovations will make possible the production of biofuels from cellulose materials (agricultural and forestry waste).

Public Transportation

According to Peter Newman (2009:108), "The biggest challenge in an age of radical resource-efficiency requirements will be a way to build fast rail systems for the scattered car-dependent cities." Unfortunately, the motor vehicle lobby represented by car, oil, and tire companies, as well as motor vehicle drivers' associations, has served to shape urban transportation and spatial patterns in terms of industries, businesses, and residential areas. Ultimately, as Stilwell (1992:35) asserts, the "urban problem of transportation starts to look like it is rooted in the power structure of the corporate capitalist economy even more than it is rooted in the spatial structure of cities."

As part of a challenge, a global movement to make inner cities car-free has emerged (see Shiva 2008:53). Sustainable transportation would entail many other measures, such as limiting the use of cars as much as possible, making them smaller and more energy efficient, and even banning four-wheel drives except in special circumstances (such as in the outback and mountainous areas) and drastically limiting air travel. Human societies need to move away from private motor vehicles as much as possible and make them smaller and more energy efficient than they generally are now. Electric cars are often offered as a more environmentally sustainable form of transport. This may be the case if they derive their power from renewable sources of energy but not necessarily if they derive their energy from coal-fired power plants. Furthermore, electric cars will not solve congestion problems and the need to build and maintain roads, which requires an enormous amount of concrete, the manufacture of which produces CO_2 emissions.

In Europe and various other places around the world, efforts to develop efficient public transport are being extended to outer suburbs and even rural areas (Mees 2010:7). Much thought is being given to the best form of public transport, such as train, tram, or bus, depending on the situation. For example, for a variety of reasons, including poor scheduling or limited routes, buses are frequently underutilized. According to Mees (2010:38), "A bus with half a dozen passengers will be no more efficient, in greenhouse terms, than if the passengers travelled in cars at average occupancies." Conversely, more fuel-efficient cars, whether electric, hybrid, or smaller, can be a problem in that viewing them as a panacea tends to absolve "policy makers of the need to make substantive changes to transport policy" (Mees 2010:41). Furthermore, reports about gains in automobile fuel efficiency should be taken with a grain of salt as they reflect improvements achieved under controlled test conditions rather than in actual everyday traffic (Mees 2010:42).

Many urban public transport systems are privately owned, although often partially subsided with government funds. Zurich has a largely publicly owned transport system that is well integrated and efficient in an urban area of medium size and medium population density (Mees 2010:129–45). Various Latin American cities, such as Curitiba in Brazil, have created rapid, low-cost, high-frequency bus systems that utilize bus-only lanes and feeder buses that link with the larger bus network (Metz 2010:161).

Copenhagen has created a 10-step program designed to achieve a low-car/high-bicycle livable city that includes the following measures:

- Converting streets into pedestrian thoroughfares
- Reducing the number of cars permitted in the city center and eliminating parking spaces at a rate of 2 to 3 percent per annum
- Converting parking lots into public spaces
- Encouraging people to live in the city center, in part to reduce car dependence
- Promoting cycling as an important form of transport
- Providing bicycles for a nominal user fee (Metz 2010:160)
- Fostering low-slung, densely spaced buildings that allow breezes to pass over them

Many cities are developing bicycle–public transport linkages. For example, "Vancouver's Translink transport authority . . . is investing in improving safety by building separated bicycle paths along roads and bridges, providing lockers for bikes at railway stations and bike racks on buses so that bikes can be taken on board" (Low et al. 2005:141).

Obviously there is a need to shift increasingly from both private motor vehicles and even buses to trains and light-rail systems in urban areas and trains between urban areas. To compete with air travel, there has been much discussion about the need for high-speed trains, although these would not be a panacea, as the

> faster the train goes, the more energy it consumes, and the more CO_2 it emits per kilometre in order to overcome the greater wind resistance.

In capitalist societies, "time is money," and this dictates rapid movement between places. Conversely, in a more leisurely paced world based on eco-socialist principles, people might find slower train travel—although faster than exists in most parts of North America and Australia—to be a time to slow down by reading, chatting with fellow passengers, enjoying the passing countryside, and even sleeping.

One form of public transport in particular needs to be rethought: air travel. Business people, politicians, celebrities, and globe-trotting tourists are not the sole frequent flyers; so are many academics who attend conferences in far-off places. For domestic air travel, there is a need to "use larger aircraft (because they are more fuel efficient) flying over fewer routes—that is, those that can generate the high occupancies that will also be required to attain low levels of fuel use per passenger-kilometre" (Gilbert and Perl 2010b:358). The growing concern about climate change has prompted discussion about the possible revival of airships that could be powered by a hydrogen-helium mixture, thus circumventing the danger of disasters such as the explosion of the German airship *Hindenburg* in 1937 (Sim 2010a:92). Airships would constitute a form of slow travel given that they travel at speeds of 150 to 200 kilometers per hour. They also can "carry large loads with one-tenth the fuel of aircraft technology" and are already being used in eco-tourism (Newman 2010:183). Domestic air travel can be replaced by high-speed intercity rail travel of the sort that already exists in Europe and Japan. The existing infrastructure for passenger rail travel in North America and Australia, as well as in many developing countries, could be improved and expanded. In terms of international travel, there is a need for larger, more fuel-efficient airplanes, even ones that are partially solar powered. As Gilbert and Perl (2010b:358) observe, transoceanic ships "could make considerable use of wind power through the use of kites or solid sails."

A more sustainable form of holidaying would entail trips much closer to home rather than to distant lands. Over the course of my seven and a half years of residence in Australia, I have found it interesting how many Australians I have met who have travelled to distant destinations in Europe, North America, and Asia but have not visited places closer to home, such as Tasmania and New Zealand. Perhaps the irony of air travel is captured by Goodall (2010:184), who observes that he is "appalled by the international conferences on climate change that involve thousands of delegates travelling many miles by air." While many may take issue with his assertion that the people who seek to convince others of the serious nature of climate change simply "must stop flying," except in emergencies, obviously there is a strong need to think about how we could network, such as via teleconferences, with colleagues and activists, particularly those in the climate movement, in our present global situation.

Green Cities

Historically cities have been an integral component of state societies and have served as political, commercial, and cultural centers. With the rise of capitalism, cities have been shaped by the owners and managers of capital to meet their financial, commercial, and industrial interests. Urban life for most city residents separates them in large measure from nature, even if the city in question is dotted with "green spaces," such as large parklands and ocean fronts. Although cities cover less than 2 percent of the Earth's land surface, reportedly "they account for some 75% of global energy demand and they produce 80% of CO_2 and greenhouse gases" (Abbate 2008:130). Cities have contributed to much of the planet's environmental degradation (e.g., water and air pollution) and loss of forests, farmland, and biodiversity. Over 50 percent of the world's population now lives in cities, and some scholars project that 75 percent of the world's population will reside in cities by 2050 (Abbate 2008:130). Whether or not we like the city, for better or worse, it is a human phenomenon and here to stay, although hopefully not in its present forms. While cities are places where the rich and powerful wield their wealth and power over the rest of us, they are also vibrant sites where intellectual and political ferment often crystallize. Furthermore, they are the "loci of much of our most insightful critiques of consumerism and its harmful ecological (and social) consequences" (Low et al. 2005:39).

Presently, cities are environmentally unsustainable and account for 75 percent of the world's fossil fuel consumption, making them the single greatest source of climate change (World Council for Renewable Energy 2005). Conversely, at least in developed countries, cities may be more environmentally sustainable if they are more relatively compact and have relatively good public transport systems as opposed to being sprawling and auto dependent. According to one study, New York City has "one of the lowest per capita CO_2 emissions of any large Western city, less than a third of the per capita US average" due to its high population density, walkability, relatively good public transport system, and relatively low heating requirements for apartment living (there is lots of room for improvement in this regard) (Hopkins 2008:29). More and more urban planners are calling for the development of "green cities" that include numerous green spaces, efficient and inexpensive public transport, energy-efficient offices and dwelling units, and vibrant community centers. Cities need to be redesigned in such a way that people live closer to their jobs, ones that make a meaningful contribution to social welfare and environmental sustainability. Magdoff and Foster (2010:27) argue that "smaller cities may be needed, with people living closer to where their food is produced and with industry more dispersed and operating at smaller scale." I have often asked myself why Australia, with its huge amount of land, has over one-third of its population concentrated in two metropolitan areas. Sydney now has well over 4 million people, and Melbourne is quickly approaching 4 million people. There have been discussions to shift the urban growth to some of the smaller cities or "country towns," but thus far these efforts have failed due to a lack of political will.

Fortunately, a new urbanism that seeks to make cities more livable and environmentally sustainable has emerged around the world and has even permeated urban planning in Australia. One of the primary sources of greenhouse gas emissions in cities is motor vehicles. In addition to having a detrimental impact on the environment, cars are very expensive modes of transport. Peter Newman and Isabella Jennings (2008:45) report, "Cities that are car dependent spend between 15 and 20 percent of their wealth just on getting around, whereas transit-oriented cities spend only 5 to 8 percent of their wealth on transport." Richard Register (2001) maintains that cities should be designed for people, not for cars. Furthermore, he proposes the notion of *pedestrian cities* in which people will not need cars and will be able to walk, cycle, or take public transportation to get around. Community gardens and family-unit gardens increasingly are viewed as part of the effort to address environmental problems on a number of fronts, including in terms of climate change. Urban eco-villages that "seek to nurture social and economic security through creating cooperative empowered communities, and local production and livelihoods" (Newman and Jennings 2008:46) have emerged in Ithaca (New York), Los Angeles (California), Adelaide (South Australia), and other cities in at least the developed world. In reality, the development of green cities constitutes a "real utopia" in that there are already efforts around the world to develop more sustainable urban environments, such as in cities as diverse as Dar es Salaam, Havana, Hanoi, Caracas, Paris, Chicago, Philadelphia, Melbourne, and Brisbane (Brown 2009:160–61). Groningen, a city of some 170,000 in the Netherlands, removed the roads in its central business district in 1992 and adopted various steps promoting bicycle transport (Korten 2001:256). Some psychologists have coined the notion of *eco-psychology*, which stresses the need for people, including urban dwellers, to have contact with the natural environment (Brown 2009:162). H. Frumkin (2001:238) observes,

> There is evidence . . . that contact with the natural world—with animals, plants, landscapes, and wilderness—may offer health benefits. Perhaps this reflects ancient learning habits, preferences, and tastes, which may be echoes of our origins as creatures of the wild. Satisfying these preferences—taking seriously our affiliation with the natural world—may be an effective way to enhance health.

Unfortunately, most of the efforts to make cities greener have benefitted the affluent rather than the poor or even ordinary working-class people. Ultimately, the development of green cities will have to be part and parcel of achieving great social equality on a global scale.

Resisting the Culture of Consumption, or What Is It That We Really Need, Materially and Otherwise?

Abraham Maslow delineates five categories of "basic needs": physiological needs, safety needs, belongingness and love needs, esteem needs, and the need for self-actualization. Marx distinguishes between "true" and "false" needs—"that is, between things that people really do need and things that they falsely believe themselves to need—and criticises capitalism for its tendency to induce false needs" (Hughes 2000:169). Obviously, all humans need to consume a certain amount of food, clothing, and shelter to sustain themselves. In modern societies humans need to engage in certain forms of *collective consumption*, such as education, health care, and public transport, which often are socialized, at least in part, because they do not result in high profits (Stilwell 1992:166–67). Capitalism has a strong predisposition to convert "needs" into "wants" through voluminous and enticing advertisements and as a compensation for alienation in the workplace and everyday social life (or lack thereof). Thus, the endless consumer items that people accumulate, particularly in developed societies and even among the affluent in developing societies, are not essential to their material well-being per se and, while contributing to capitalist profit making, are environmentally unsustainable. Herbert Marcuse (1991), a leading figure of the Frankfurt School, in *One Dimensional Man*, argued that capitalism not only exploited workers for their labor power but assuaged their alienation by seducing them to participate in a consumerist society. Indeed, Thich Nhat Hanh, a renowned Zen Buddhist master and not an eco-socialist, captured in an interview in 2010 the interconnectedness of the treadmill of production and consumption, environmental degradation, and alienation. He noted that "the situation the Earth is in today has been created by unmindful production and unmindful consumption. We consume to forget our worries and our anxieties. Tranquilizing ourselves with over-consumption is not the way" (quoted in Gilding 2011:214).

While sustainable lifestyles and green consumerism are often regarded as "shallow" forms of environmentalism, they can function as part and parcel of resistance to the capitalist treadmill of production and consumption (Spaargaren and Mol 2008:357). Giddens (2009:65), hardly a radical, maintains, "We can legitimately talk of *over-development* as a possibility in the affluent societies." Indeed, the more affluent sectors of developing societies are also exhibiting overdevelopment. Conversely, we can also speak of underdevelopment as existing among the abject poor in both developed societies and developing societies. Sim (2010b:179) proposes a "politics and an economics of enough."

Stilwell (2000:56) argues that while "capitalism has certainly proven to be impressive . . . as a means of producing a vast array of commodities," many people have come to recognize that "our principal sources of satisfaction—personal security, social recognition and interesting work—are often destroyed in the preoccupation and consumption." Furthermore, as I have noted, the capitalist treadmill of production and consumption constitutes a major source of environmental degradation, including particularly climate change, which is a global phenomenon, not merely a localized one.

Ted Trainer (1989, 1995) provides several proposals that could serve as transition points between the existing capitalist system worldwide and an alternative socially just and environmentally sustainable social system. He calls for "appropriate development" for both "rich" and "poor" countries. In terms of the former, this would entail an enormous reduction in consumerism (Trainer 1989:196) and essentially reversion to a "zero-growth economy" in which societies "will work hard at reducing the amount of producing and consuming going on" (Trainer 1995:108). In terms of the latter, this would include a focus on local economic self-sufficiency; the utilization of "low, intermediate, and alternative technologies processing locally available resources"; and a commitment to environmental sustainability (Trainer 1989:199–201). Furthermore, in the context of a "simpler way," people could shift from production and consumption of material items to less resource- and energy-intensive activities, such as the arts, exercise, meditation, and even alternative therapies such as massage.

Trainer (1998:8) argues that the "fundamental cause of the accelerating destruction of the global ecosystem is overproduction and overconsumption" of material goods. He argues that developed societies such as Australia exceed sustainable levels of production and consumption (Trainer 1998:8). Trainer (1998:10) asserts that Australians "can only live as affluently as we do because we are taking and using up most of the scarce resources and preventing most of the world's people from acquiring anything like a fair share." Trainer (1991:124) has for a long time called for a "conserver society" committed to great equality in income and wealth—one with "much lower than present rates of per capita income and resource use" that contributes to the creation of a "world order that is peaceful, just, ecologically sustainable and in which inequality and poverty have ceased to exist." His notion of a "simpler way" incorporates the following principles: (1) far simpler material living standards; (2) high levels of self-sufficiency within households, nationally and especially within neighborhoods and towns; (3) relatively little long-distance trading and transport; (4) small-scale economies in which most of things we need are produced by local labor from local resources; (5) cooperative and participatory local systems; (6) an alternative economy that does not entail growth and requires far less work and production and consumption than the present one; (6) a commitment to human rights and social justice, particularly with regard to developing countries; and (7) a radically different culture. Trainer (1996:143) argues that "consumer-capitalist society" cannot be repaired and proposes a blending together of social planning, cooperative economic endeavors, and even a small "enterprise sector."

From an eco-socialist perspective, Magdoff and Foster (2010:26) argue the following:

> An economic system that is democratic, reasonably egalitarian, and able to set limits on consumption will undoubtedly mean that people will live at a significantly lower level of consumption than what is sometimes referred to in a wealthy country as a "middle class" lifestyle (which has never been universalized even in those societies). A simpler way of life, although "poorer" in gadgets and ultra-large luxury homes, can be richer culturally and in reconnecting with other people and nature, with people working the shorter hours needed to provide life's essentials. A large number of jobs in the wealthy capitalist countries are non-productive and can be eliminated, indicating that the workweek can be considerably shortened in a more rationally organized economy.

In a similar vein, Ana Isla (2009:202) notes that "materialist ecofeminists, like Maria Mies, Veronika Benneholdt-Thomsen, or Vandana Shiva, advocate alternative consumption norms based on 'enoughness,' 'sufficiency,' 'subsistence economies,' and 'gift economies.'"

In reality, most people in developed societies will need to scale back their consumption of material goods. In one of his popular online commentaries, Immanuel Wallerstein (2007), the principal architect of world systems theory, delineates three overarching obstacles to overcoming climate change: (1) the "interests of producers/entrepreneurs," who act as the purveyors of the capitalist treadmill of production and consumption; (2) the "interests of less wealthy nations," like China and India, which are emulating the developed countries; and (3) the "attitudes of you and me." While it is important not to place the burden of climate change mitigation strictly on individuals by urging them merely to become "green consumers," as Wallerstein asserts, climate change mitigation starts at the individual level, particularly for the affluent people in both developed and developing countries.

Jonathan Neale (2008:45) cautions climate activists not to talk about sacrifice to ordinary people, and on this score I am fully in agreement with his assertion that "they will find themselves without the support of ordinary people." My comments of resisting the culture of consumption are directed primarily to the affluent, even the affluent in the working class, who turn to consumerism as a compensation for alienation in the workplace and in everyday social life in modern capitalist societies. As Neale (2008:46) so astutely asserts, it is imperative that the "idea of social justice and sacrifice" be a central theme in the climate movement and that climate activists be part of an effort to "mobilise the people of the world to survive, share, help each other and make the established powers of the world sacrifice."

Paul Lafargue (1999:3), in *The Right to Be Lazy*, advises, "Let us be lazy in everything, except in loving and drinking, except in being lazy." Besides the issue of whether "laziness" is a virtue, it is incompatible with capitalist exploitation and thus constitutes an "active assertion of an alternative practice" and thus an "effective form of resistance" (Holloway 2002:24).

Telecommuting can contribute to decline in work trips (Schaefer et al. 2009:57). Virtual reality technology can immerse workers in virtual travel. As mentioned, long-distance travel for business, political gatherings, and conferences of all sorts can be greatly reduced by video conferencing (Black 2010:219).

Sustainable Agriculture and Forestry

One important climate change mitigation would be a strong shift in food production away from a heavy reliance on meat, particularly livestock, to organic farming, vegetarianism, and even veganism. Small-scale organic farming tends to be more fuel efficient than industrial agriculture, which relies heavily on oil and artificial fertilizers and pesticides. None other than the

United Nations Environment Programme (2010) has called for a global shift toward a vegan diet. All farming requires water, but livestock production requires much more water than does growing crops. For example, whereas producing 1 kilogram of wheat requires 1,000 liters of water, 1 kilogram of beef requires 100,000 liters of water (Godrej 2006:67). Even in those instances where animal production continued, methane from animal digestion could be reduced through practices such as vaccinations and chemical inhibitors and changes in tillage practice and vegetation cover. It may also be possible to reduce N_2O emissions from fertilizers through fertilizer management, soil and water management, and fertilizer additives (Rickards and Tucker 2009:93). Agricultural climate change strategies include (1) enriching soil carbon, (2) creating high-carbon cropping systems, (3) promoting climate-friendly livestock production systems, (4) protecting existing carbon stores in natural forests and grasslands, and (5) restoring vegetation in degraded areas (Scherr and Sthapit 2009:33). Aquaculture is often proposed as a solution to declining fish stocks in the oceans, but unfortunately farmed fish have to be fed many times their body weight in wild fish meal.

Due to severe shortages of fuel, chemicals, and feed during the Special Period following the collapse of the Soviet Union, Cuba was forced to adopt "ecological agriculture," including permaculture, a form of organic farming initially developed by Bill Mollison, an Australian writer. According to Wall (2010b:45),

> The method uses tree crops and mulches to avoid the need for labor-intensive digging. Another principle of permaculture is companion planting, where inter-cropping of different plants is used to reduce pests and increase fertility. Composting is vital. Worm bins are used to turn waste into natural fertilizers and mulches.

Cuba has been shifting away from monoculture, which in the past relied heavily on sugarcane and tobacco production, to the mixed production of fruits, vegetables, grains, livestock, and fish (Levins 2005:23). Even oxen complement tractors, with horses and other animals consuming weeds. Within the domain of ecological agriculture, Cuba has promoted urban agriculture, which entails some 30,000 hectares producing more than 3 million tons of fresh vegetables per year. According to Richard Levins (2005:22), urban agriculture

> provides abundant, diverse fresh vegetables throughout the year for consumers. This has transformed the Cuban diet in the communities, schools and workplaces and encouraged the spread of vegetarian restaurants. It lowers the costs of transportation and storage by selling directly to consumers. It provides employment for some 300,000 people at a time when capital is not available to invest in more industrial employment.

Residents of Havana plant vegetables and fruits on their rooftops and what had been vacant lots. With a shortage of oil resulting from the collapse of the Soviet Union, one of the strategies that Cuba adopted was replacing some 40,000 tractors with some 385,000 oxen (Shiva 2008:76). In her film *The Power of Community: How Cuba Survived Peak Oil* released in 2006, Megan Quinn-Bachman, the outreach director of the Arthur Morgan Institute for Community Solutions in Yellow Springs, Ohio, documented how Cubans have adopted *relocalization* as a strategy for addressing energy shortages. Indeed, there has been some relocation of Cubans, partly voluntarily and partly government induced, from cities to rural areas, a trend that is completely the reverse of what is occurring in most developing countries (Heinberg 2010:57).

A slow food movement has emerged around the world, particularly in developed countries, which includes not only food preparation at home but reliance on locally produced foods. According to Carl Honore (2004), the slow food movement is part and parcel of a larger global movement calling for slow cities, slow medicine, slow sex, slow work, slow leisure, and slow child rearing.

Conclusion

The transitional steps that I have delineated constitute a loose blueprint for shifting human societies or countries toward democratic eco-socialism and climate change mitigation, but it is important to note that both of these phenomena cannot be achieved in one country alone. The process has to be global, and the ways in which it will be carried out will vary from country to country. On the issue of climate governance, Camilleri and Falk (2010:287) maintain that any effective response to global warming or climate change requires the "involvement of many institutions, at many levels of spatial organisation, across many societies, and over the long term." Bearing in mind their further assertion that "global warming has paradigmatically changed the scope of human governance," world systems theorists argue that this will eventually require a socialist world government that is interwoven with a model of global democracy that is "premised on the democratic self-determination of nations" (Boswell and Chase-Dunn 2000:176).

The shift from global capitalism to global democratic eco-socialism will inevitably entail the elimination of many jobs that contribute to greenhouse gas emissions and the treadmill of production and consumption, such as ones in the fossil fuel industries (oil and natural gas extraction, coal mining, manufacture of motor vehicles with internal combustion engines, etc.) and ones in marketing and sales. Conversely, the transitional reforms and ultimately the creation of global democratic eco-socialism will entail the creation of new jobs, in the manufacture of renewable energy facilities and public transport systems. Furthermore, new service jobs in health care, education, recreation, the arts, and environmental restoration, which serve social needs and are often restricted in a capitalist system because they are viewed as cutting into profits or not creating profits, could be created. Finally, under democratic eco-socialism, there would not be the gross inequities in hours that people work, with many "overworked" and others unemployed or underemployed.

8

Grassroots Responses to Climate Change: Internationally, Nationally, and Locally

Ordinary people around the world are responding to anthropogenic climate change in a wide variety of ways, in some cases by denying or downplaying its existence, in many cases by admitting its existence but asserting that responding to it is not in their sphere of action, and finally by becoming involved in a burgeoning and disparate climate movement that exists at international, national, and local levels. In this chapter I examine three scenarios. The first of these draws from an ethnographic study done by sociologist Kari Marie Norgaard (2011) in which she examines a rural community in western Norway, particularly during the warm winter of 2001 and 2002, and how highly educated people who are situated in a society committed to egalitarian and social justice issues but who are also the beneficiaries of global capitalism, in large part due to the oil and natural gas boom, have managed the contradictions in their daily lives. The second scenario draws from a study conducted by sociologist Christine Shearer (2011) on the Inupiat community of Kivalina in Alaska, which must be relocated so that its residents can avoid the impact of the rising sea and storm surges. The community has challenged the fossil fuel industry, albeit unsuccessfully thus far, with a lawsuit for not acting in a more responsible manner on climate change mitigation. The final and much longer scenario focuses on the climate movement both internationally and more specifically in the United States and Australia.

Coping Emotionally with Climate Change: The Case of a Rural Norwegian Community

Norgaard (2011) conducted ethnographic research in Bygdaby (a pseudonym pronounced "big-DAH-bee"), a town of some 10,000 to 14,000 people situated in a mountainous region of western Norway at the dawn of the present century. The community consists in large part of single-family farms but also has commercial, communication, education, medical, and tourist enterprises. In the fall of 2000 and winter of 2001, the community experienced unusually warm weather.

> November brought severe flooding across the entire region. By early December, it was established that the weather was measurably warmer than usual. The local newspaper reported that October, November, and December were respectively 4.0, 5.0, and 1.5 degrees [Fahrenheit] warmer than the 30-year average. As of January 2001, the winter of 2000 for Norway was recorded as the second warmest in the past 130 years. . . . In the town of Bygdaby . . . the first snowfall did not come until late January—some two months later than usual. (Norgaard 2011:xiii)

In contrast to large pockets of climate change skepticism in the United States and Australia, in Bygdaby Norgaard encountered very few climate skeptics, although some of her informants were unclear about the mechanisms involved in anthropogenic climate change, such as carbon dioxide emissions. Despite this, she found the vast majority of her informants were not taking action, particularly collective action, on climate change. A climate movement that has emerged in other places, including very likely Oslo, was nowhere to be seen in the town and its environs, despite the fact that the community residents, like Norwegians in general, are highly aware of their environment and thus in a "special position to notice climate change and to have their identity affected by it" (Norgaard 2011:20).

In contrast to many rural communities in the United States and to a lesser degree Australia, Bygdaby has no poor people, a "reflection of the fact that Norway is a social democracy with one of the most comprehensive welfare states in the world" (Norgaard 2011:15). Furthermore, many community residents are politically engaged and express their political engagement by attending local party meetings; they even participate in demonstrations of various sorts, including on May 1, Labor Day, and opposition to the European Union, genetically modified foods, and racism. Bygdaby has eight active political parties, including the Labor, Center, Socialist Left, Progressive, Red-Electoral Alliance, Christian Democratic, Liberal, and Conservative parties. Despite a high level of political activism there, Norgaard (2011:43) notes that "climate change was never discussed in any of the city council or specific Labor Party strategy meetings I attended or even in the meetings of the municipal subgroup on culture and environment." Informants tended to view climate change as more of a national or international issue that could not be addressed at the community level in any significant way. Teachers noted that students often resisted discussion about environmental issues and climate changes. When Norgaard (2011:55) mentioned climate change in her interviews, she "noticed that it often killed the conversation," despite an initial expression of concern. There was a tendency to move the discussion onto more familiar topics, such as moral deterioration. On the whole, community residents tended to feel helpless and powerless in the face of climate change, the effects of which they were experiencing firsthand.

While Norway is a more progressive country on the whole than either the United States or Australia, it very much is part of the global North in that it has become the largest oil producer in Europe and the world's fifth-largest oil exporter, and it derives one-third of its national revenue from oil. Although as a signatory of the Kyoto Protocol, Norway promised to limit its greenhouse gas emissions to a maximum of 1 percent of 1990 levels, its emissions increased from 35.2 million tons of carbon dioxide in 1990 to 53.8 million tons in 2008 (Norgaard 2011:71). Furthermore, as of 2008, the oil and gas industries accounted for 26.6 percent of Norway's CO_2 emissions. Given these contradictions, Norgaard (2011:70) maintains that the "people I spoke with in Bygdaby played a critical role in legitimizing the status quo by not talking about global warming even in the face of late winter snow and a lake that never froze." She asserts that citizens of developed countries who do not respond to the reality of climate change benefit from their denial economically as well as by "avoiding the emotional and psychological entanglement and conflicts that may arising from knowing that one is doing the 'wrong thing'" (Norgaard 2011:72). Norwegians tend to view themselves as a people committed to egalitarianism and environmental responsibility, but for many climate change disrupts their notions of moral sensibility. In actual fact, however, most Bygdaby residents and other Norwegians are not much different from most people in developed societies and very likely affluent people in developing countries.

Probably most of us in developed countries are wittingly or unwittingly complicit in contributing to climate change in our lives, and we are constantly bombarded with pressure from corporations to consume this or other product, whether we actually need it or not. Indeed, Hornborg (2001:25) observes that with the collapse of the protest movements of the 1960s and early 1970s in the developed world, many people probably sensed that "their affluence was based on the impoverishment of the South and the global environment [which they found] unbearable and thus impossible to accept." In my conversations with middle-class and even relatively affluent Australians, I sense the cognitive dissonance of which Hornborg speaks. While many of these people argue that they engage in a variety of activities that are environmentally friendly, such as using canvas shopping bags, recycling, and placing solar panels on the roofs of their houses, responding to climate change will ultimately take more than individual acts of "green consumerism" and will entail collective action of one sort or another, something that we see manifested among the Inupiat people of Kivalina, Alaska, and in the climate movement in its various manifestations around the world.

An Indigenous Response to Climate Change: The Case of Kivalina in Alaska

Along with Shishmaref, an Inupiat village of some 600 residents on Sarichef Island in northwestern Alaska, Kivalina, a village of some 400 residents situated about 120 miles north of the Arctic Circle, constitutes yet another example of the adverse impact of climate change on indigenous peoples around the world. Shearer (2011:15) reports,

> The tiny village sits on a thin strip of land, a quarter mile across at its widest point, and is sandwiched between the Chuckchi Sea and Kivalina Lagoon, at the mouths of the Wulik and Kivalina Rivers. The sea's waves eat at the shore from the west and the water slowly undercuts it from the east. The worst threats, however, come from the storms, which have swallowed as much as seventy feet in one downpour. Kivalina traditionally enjoyed protection from storms by sea ice formations, which surrounded and hardened the coastline. For the past three decades, however, the ice has formed later in the year and melted earlier, leaving the shore vulnerable to quick and dangerous erosion from storms. With a maximum elevation of only ten feet above sea level, Kivalina residents have lived in constant alert.

Kivalina residents, who still retain a largely foraging lifestyle, first observed the erosion of their island in the 1950s and voted to relocate in 1992; they selected a new site in 1998 but quickly discovered that no government agency existed that could assist them with their proposed relocation (Shearer 2011:102). The community has thus far had little success with its relocation efforts, despite the fact that its members find themselves in increasing danger.

In February 2008, Kivalina, along with various indigenous rights and environmental justice organizations, filed a suit against 24 oil, electricity, and coal companies for contributing to the village's erosion by contributing to greenhouse gas emissions and climate change and systematically fostering climate change denial. The village asked the fossil fuel industry to pay for its relocation costs, estimated at $100 million to $400 million. Unfortunately, following highly complicated and drawn-out legal wrangling, Kivalina's claim was dismissed, requiring the community members to seek "other means to protect themselves and their homeland" (Shearer 2011:124). Unfortunately, class action suits are generally very difficult and often very expensive within the context of the US legal system. The case of Kivalina is yet one more tragic example of how various peoples around the world who have contributed the least to climate change are suffering, and will continue to suffer, the most from its ravages.

The Climate Movement: Internationally and Nationally

The climate movement, both internationally and nationally, is a broad and disparate phenomenon that in part draws upon earlier movements, particularly the environmental movement but also the global justice or anti–corporate globalization movement, the indigenous rights movement, and the more progressive segments of the labor movement. Figure 8.1 presents a schematic depiction of the various tendencies in the international climate movement:

Athanasiou and Baer (2002:128) see commonalities between the antiglobalization and climate movements in that "both strain to rework the terms of global finance" and "both focus on the issue of ecological sustainability." Many climate groups in North America, Europe, and Australia tend to focus on ecological modernization as their primary strategy for achieving climate change mitigation, thus either ignoring or downplaying social justice issues. Julian Agyeman, Harriet Bulkeley, and Aditya Nochur (2007:144) address this shortcoming by arguing,

> Just as environmental justice groups challenged the mainstream environmental movement to consider environmental justice issues in the early 1990s, the climate movement today must develop a climate justice analysis. There is immense potential to support the work of and build coalitions with groups

Table 8.1. Tendencies in the International Climate Movement

Green Democratic Tendency
- Tends to be more pronounced in developed societies
- Seeks to regulate capitalism
- Places strong emphasis on ecological modernization
- Is rather muted on social-equity and justice issues

Radical and Anticapitalist Tendency
- Is eco-socialist or eco-anarchist
- May be most pronounced in developing societies but does exist in developed countries
- Views ecological modernization as important but insufficient to contain climate change
- Calls for transcendence of global capitalism with an alternative social system committed to social parity and justice and environmental sustainability

In-Between Tendency
- Recognizes social justice issues but is not explicitly anticapitalist

> working on climate justice issues. . . . Communities that are most affected by climate change have the most critical stake in mitigating the problems. Because these groups tend to be excluded from policy debates, we must take additional steps to ensure that their voices are heard.

The international climate movement appears to have started in 1989 with the formation of the Climate Action Network (CAN), which by mid-2008 had grown into an alliance of some 430 NGOs (Camilleri and Falk 2010:309). Unfortunately, as an umbrella group CAN has had a mixed record. It lobbied heavily in support of the Kyoto Protocol and still supports emissions trading schemes or their passage, such as in Australia. Some of its leaders have had connections with carbon-trading companies, including board member "Jennifer Morgan of the Worldwide Fund for Nature, who took leave for two years to direct work on Climate and Emergency Security at carbon-trading firm E3G," and "Kate Hampton, formerly of Friends of the Earth, who joined Climate Change Capital as head of policy while simultaneously advising the EU on energy and the environment, working for the California Environmental Protection Agency, and acting as president of

International Carbon Investors and Services" (Bond 2009:215). Stop Climate Chaos formed in the United Kingdom in September 2005 as a coalition of over 70 NGOs and groups, including Greenpeace, Islamic Relief, Friends of the Earth, World Wildlife Fund–UK, Youth against Climate Change, and the Royal Society for the Protection of Birds (Castells 2009:323–24). The climate movement both at the international level and within specific countries is quite divided, with some of its segments trying to work within the parameters of global capitalism and others challenging it head on. Friends of the Earth, an NGO with national organizations in 70 countries and some 5,000 local groups, has engaged in a major campaign promoting climate justice, particularly for peoples in developing countries who are being adversely impacted by climate change. According to Manuel Castells (2009:325), "The internet has played an increasingly important role in the global movement to prevent global warming."

One contingent of the climate movement emphasizes a "leave it in the ground" approach that seeks to halt the mining and shipment of coal. The International Rising Tide Network in particular has been a key player in this campaign and sponsored Fossil Fuels '08 and '09 rallies in the United Kingdom, United States, Australia, and South Africa (Cooke 2010:425). Climate camps directed at coal mines, coal ports, and coal-fired power plants have occurred in the United States, Australia, Germany, Canada, New Zealand, France, Belgium, Holland, Denmark, India, Ireland, and the Ukraine.

Some climate action groups (CAGs) have begun to target biofuel operations around the world. For instance, over 200 climate campers blockaded the world's largest agrodiesel refinery in Hamburg, Germany (www.viacampesina.org; Abramsky 2010). Maoist guerrillas raided a biofuel plant in the central Philippines that the country has been promoting to reduce its reliance on oil.

The Transition Town movement has emerged as a segment of the climate movement in that its adherents are seeking to have local neighborhoods transition into climate-friendly communities (Lohan 2009). Rob Hopkins (2008), a permaculture teacher, has served as the charismatic leader of the movement and founded the Transition Town Totnes, in Totnes, Devon, England, in September

2006 (Scott-Cato and Hillier 2010:874). The movement seeks to "create much more self-contained communities that have successfully weaned themselves off scarce fossil fuels and dramatically reduced their greenhouse gas emissions" (Vaze 2009:326). The Transition Town movement seeks to move away from globalized distribution systems to localized ones as a challenge to peak oil and climate change and to rejuvenate local economies.

As of February 2009, there were reportedly 134 officially registered Transition Towns (Scott-Cato and Hillier 2010:874; www.transitions.org). Molly Scott-Cato and Jean Hillier (2010) conducted ethnographic research on several Transition Towns in England. Transition Town Stroud engages in a "process called 'Learning from the South'—both a reminder that the sustainable future does not mean cutting ourselves off from the poorer world and our responsibility for others, and in recognition of the ability many in the poorer world demonstrate to live happier and more sustainable lives than our own" (Scott-Cato and Hillier 2010:876). Many Transition Towns have shown the film *The Power of Community: How Cuba Survived Peak Oil* (2006), which describes Cuba's efforts to become more self-sufficient in agriculture in the aftermath of the collapse of the Soviet Union, when it lost a major source of its oil supplies. Transition Towns' working groups examine issues such as energy, health, education, food, clothing, subsistence, and community building. Transitional initiatives often start with food projects ranging from "community-supported agriculture to urban fruit and nut tree plantings, from community gardens to re-skilling workshops around food production" (Hopkins 2010:449).

In some ways the Transition Town movement resembles the counterculture movement of the late 1960s and early 1970s in that "Transition Towns see their role as creating sustainable livelihoods outside the formal economy, through self-provisioning and the creation of alternative currencies" (Scott-Cato and Hillier 2010:882). In large part, however, the Transition Town movement has been a phenomenon of the developed world with its initiatives presently based primarily in the Anglophile countries, particularly the United Kingdom but also the United States, Australia, Canada, and New Zealand (Hopkins 2010:447).

Ironically, while New Zealand has about one-fifth the population of Australia (4.2 million versus 22 million people), the Transition Town movement has progressed further in the former than the latter, perhaps in part because of the greater compactness of New Zealand society (Hopkins 2008:218–36). While sympathetic toward the Transition Town movement, Ted Trainer (Trainer 2009), an Australian eco-anarchist, expresses reservations about it:

> There is the danger that it will only be a Not-In-My Backyard phenomenon, that it will be about towns trying to insulate themselves from the coming time of scarcities and troubles. This is a quite different goal from working to replace consumer-capitalist society. It is not much good if your own town bakes its own bread or even generates much of its own electricity, while it goes on importing hardware and appliances produced in China and taking holidays abroad. It will still indirectly be using considerable amounts of coal and oil in the goods it imports.

Furthermore, Hopkins (2008), in his extensive discussion of the political and ecological premises of the Transition Town concept, tends to downplay how global capitalism contributes to climate change and social disparities within and between nation-states.

Agyeman, Bulkeley, and Nochur (2007:136) include a broad array of groups under the rubric of what they term the *international climate justice movement*. These include various think tanks, such as the Centre for Science and Environment in India; the Indian Climate Justice Forum; Rising Tide, an international network with groups in Europe, North America, and Australia; religious groups such as the European Christian Environment Network; and indigenous groups such as the US-based Indigenous Environment Network. Bill McKibben, an American environmentalist, and others formed 350.org as an effort to create a global climate movement. In reality, many environmental NGOs, such as Greenpeace, Friends of the Earth, Oxfam, Christian Aid, and Tearfund, have become part and parcel of a climate movement that actually includes many other types of groups, including socialist and anarchist ones and student environmental collectives. Conversely, some environmental groups that have become involved in the climate movement, such as the Worldwide Fund for Nature, the Environmental Defense Fund, Greenpeace, and the Sierra Club, reportedly have strong corporate connections (Bond 2010:24).

The US Climate Movement

Mary Lou Finley (2007:46) has argued that the "climate movement in the United States is in the midst of movement take-off." The Green House Network aims to "multiply leadership supporting the clean-energy revolution that we need to stop global warming" and has conducted training workshops since 1999 to achieve this goal in partnership with many regional bodies, including Clean Air–Cool Planet in New England, the Massachusetts Climate Action Network, the Blue Water Network, Redefining Progress in Oakland, the Grand Canyon Trust in the Southwest, the Environmental Law and Policy Center in the Midwest, and Climate Solutions in the Pacific Northwest (Goodstein 2007:159). Various environmental NGOs, such as the Environmental Defense Fund and the Nature Conversancy, fall into the green social democratic wing of the US climate movement. Indeed, the Environmental Defense Fund reportedly "partnered with BP in the installation of its in-house emissions trading scheme in 1997, contributing to the corporation's post-GCC [Global Climate Coalition] green image."

In April 2001, Redefining Progress, an Oakland, California–based think tank, formed the Environmental Justice and Climate Change Initiative, a network consisting of 28 US environmental justice, climate justice, religious, policy, and advocacy groups (Agyeman, Bulkeley, and Nochur 2007:139). The initiative released its 10 principles for just climate change policies at the 2002 World Summit on Sustainable Development:

- Stop cooking the planet.
- Protect and empower vulnerable individuals and communities.
- Ensure just transition for workers and communities.
- Require community participation.

- Global problems need global solutions.
- The U. S. must lead.
- Stop exploration for fossil fuels.
- Monitor domestic and international carbon markets.
- Caution in the face of uncertainty.
- Protect future generations.

Other US organizations working for climate justice include the Indigenous Environmental Network, the Global Justice Ecology Project, and Rising Tide. These groups and others in autumn 2008 formed Mobilization for Climate Justice (actforclimatejustice.org), an organization that seeks to serve as a link between the US and international climate movements. The Massachusetts Climate Action Network assists people in conducting household greenhouse gas inventories and promoting clean energy projects (Isham and Waage 2007:16). Environmental Defense, the Natural Resources Defense Council, and the Nature Conservancy played a pivotal role in the creation of the climate change bill in June 2009 (Tokar 2009a). Al Gore established the Alliance for Climate Protection, which launched a campaign in April 2008 to inform the public about the seriousness of climate change and the role of human activities in contributing to it (Castells 2009:324).

On March 2, 2009, about 4,000 people assembled at the Capitol Power Plant in Washington, DC, with over half of these individuals committing civil disobedience (Russell 2009). The protesters included numerous notables, such as James Hansen, Gus Speth (a former environmental advisor to Jimmy Carter), Wendell Berry, Vandana Shiva, Bill McKibben, Congressperson Eleanor Holmes Norton from the District of Columbia, and actress Daryl Hannah. Capitol Climate Action states that the rally aimed to highlight three overarching issues: (1) the role of coal as a driver of climate change, (2) the urgency of the new Obama administration's not delaying in acting on climate change, and (3) the need for a massive "climate justice movement" built around a program of nonviolent direct action and civil disobedience.

Mobilization for Climate Justice is a North American–based network of organizations and individuals that espouse nonviolent direct action and public education in order to counteract climate change (Tokar 2009). Many indigenous groups that operate under the umbrella of the Indigenous Environment Network and Native Alaskans have created a website (nativeknowledge.org) in order to communicate their local knowledge about climate change (Johansen 2006:280). The Cool Community campaign urges individuals belonging to local organizations to "reduce their carbon footprint by 25 percent through participation in the Low Carbon Diet" (Gershon 2009:343).

Some evangelical Christians in the United States have become part of the climate movement. Richard Cizik, a National Association of Evangelicals lobbyist, adopted a stance that he called "creation care" around 2002; he cited various biblical verses to support his contention that Christians must act as good stewards of the Earth (Black and Weisel 2010:109). In 2006, he played a key role in the creation of the Evangelical Climate Initiative.

The Australian Climate Movement

I highlight the Australian climate movement in large part because I began conducting observations and became involved in it, particularly in Melbourne, in early 2008 (Baer 2009a). Since that time, I have acted as a scholar-activist within the movement by attending climate rallies, climate camps, climate action conferences, and meetings in Victoria, New South Wales, South Australia, and Canberra or the Australian National Territory. I conducted workshops at the 2007 and 2011 Climate Change/ Social Change conferences sponsored by the Socialist Alliance (an organization that I joined in 2009) in Sydney and Melbourne, respectively, where I had opportunities both to listen to and to interact with John Bellamy Foster. In 2009 and 2010, I served on the facilitation committee of the Climate Emergency Network (CEN) based in Melbourne and attended numerous climate action meetings and events in the Climate Action Centre at the Victorian Trades Hall in Melbourne. Periodically, I have attended meetings of Climate Action Moreland, a local climate action group situated in inner northern Melbourne, where I reside.

In some ways, my research as a climate activist parallels that of other anthropologists who have functioned as scholar-activists or partisan-observers in other social movements. For example, Jeffrey S. Juris (2008) wrote a fascinating, penetrating, and innovative ethnography of networking practices in the anti–corporate globalization movement. He blends "multisited ethnography" into a research style that he terms *militant ethnography*. Juris spent 14 months conducting research among Movimiento de Resistencia Global activists in Barcelona, Spain, from June 2001 to September 2002. He participated in numerous other anti–corporate globalization movement mass actions in various cities, including Seattle, Genoa, Brussels, and Madrid, and attended the first meeting of the World Social Forum in Porto Alegre, Brazil, in early 2001. More recently, David Graeber (2009) has written an ethnography of his involvement in the organizing of and events that led to the protest against the Summit of the Americas in Quebec City in 2001. In addition to having informal conversations with activists in coffee shops and attending spokescouncil meetings, he attended street actions at which police threw tear gas at the protestors.

The Topsy-Turvy Nature of Australian Climate Politics

Australian climate politics has come to encompass numerous actors, including corporations, the federal and various state governments, the opposition to various governments, the Australian Labor Party (ALP), the Coalition (the Liberal and National parties), the Green Party, the Family First Party, the Commonwealth Scientific and Industrial Research Organisation (CSIRO), the Australian Government Bureau of Meteorology, academic research centers, progressive and conservative think tanks, public intellectuals, labor unions, and, last but not least, a burgeoning climate movement (Baer 2009b). For some time, certain segments of Australian society—ranging from politicians, such as former Australian Labor Party prime minister Bob Hawke and leading members of the Green Party, to grassroots environmental groups and prominent climate scientists—have expressed concern about the impact of climate change on the environment and human societies, including Australia (Falk and Brownlow 1989). In 1995, the Labor government threatened industry with a carbon tax, which prompted the latter to challenge the Greenhouse Challenge Program, a voluntary program accepted by John Howard, who became the Coalition prime minister in 1996. During much of Howard's long tenure as prime minister, he kept Australians waiting for a concerted effort to address climate change by allowing a small "cabal" of lobbyists, self-described as the "greenhouse mafia" and consisting of executives from the coal, electricity, oil, cement, and aluminum industries, a pivotal role in formulating the government's climate policies (Hamilton 2007). Guy Pearse (2007) argues that the Howard government allowed its climate policy to be shaped by Australia's biggest polluters and lobbyists in Canberra, the federal capital. Many Australians accepted the Howard government's ongoing insistence that the climate science was still indefinite both in terms of the seriousness of climate change and whether the sources were primarily natural or anthropogenic.

A growing number of Australians, however, were unwilling to adopt a wait-and-see attitude toward climate change. Within this group were the majority of climate scientists working in universities and for CSIRO and the Australian Government Bureau of Meteorology, as well as environmentalists and others who were concerned about the drought that had impacted particularly southeastern Australia for nearly a decade, ravaging bushfires in New South Wales and the Australian Capital Territory during summer 2003, and the ongoing bleaching of the Great Barrier Reef. In frustration at the lack of action on the part of the Howard government, around 2003 a growing number of Australians began to form and join local grassroots climate action groups. Due to widespread pressures to take the threat of climate change more seriously, Howard commissioned a task force, which recommended that Australia adopt an emissions trading scheme (ETS), one that would be implemented in 2011 and would set a low carbon price with deeper emissions cuts being required later (Murphy 2008:1).

In his campaign for the prime ministership, Kevin Rudd, then the Labor Leader of the Opposition, promised Australians that his government would ratify the Kyoto Protocol, which it did after assuming power in late 2007. The Rudd government also promised to fast-forward the implementation of an ETS, which came to be known as the Carbon Pollution Reduction Scheme (CPRS), to 2010 and to implement various other mitigation strategies. The Rudd government also created a Department of Climate Change and designated Penny Wong, a senator from South Australia, as its minister.

On September 30, 2008, a long-awaited review headed up by Ross Garnaut (2008), a conventional economist who has served as an economic advisor to ALP governments starting with the Hawke government in the early 1980s, was released. It proposed two sets of targets: (1) one that sought to be part of a global strategy to limit greenhouse gases at 450 CO_2 equivalent ppm in which Australia would commit itself to a 25 percent reduction of 2000 levels by 2020 and 90 percent by 2050; and (2) one that accepted a limit of 550 CO_2 equivalent ppm as "more realistic in terms of what other countries are likely to do," thereby having Australia aim for a 10 percent reduction of 2000 levels by 2020 and an 80 percent reduction by 2050. The Garnaut Review called for an ETS and the development of "clean coal" technology but was quite weak in its call for development of wind and solar energy resources and a shift from a reliance on cars to mass transport. Furthermore, it accepted the premise that the Australian economy requires ongoing growth and did not address the impact that such growth per se would have in terms of greenhouse gas emissions.

Despite the conservative tone of the Garnaut Review, the Rudd government proposed an even more conservative climate policy in the form of the Carbon Pollution Reduction Scheme, which it presented initially in the form of a green paper in July 2008 and a white paper in December 2008. The Rudd government committed itself by 2020 to "reduce Australia's carbon pollution by up to 15 per cent below 2000 levels in the context of a global agreement where major economies agree to substantially restrain carbon pollution and advanced economies take on reductions comparable to Australia" and further to an "unconditional 5 per cent reduction in carbon pollution below 2000 levels by 2020" (Australian Government 2008: iv). While the Rudd government recognized that its targets were more modest than those the European Union has proposed for 2020, it asserted that its target range translated into a "34–41 per cent reduction in the per capita emissions of every Australian" given that the country's populations is "projected to grow by around 45 per cent over the period 1990–2020," where the European Union's "target range translates into 24 to 34 percent in reductions for each European" given that its population is "projected to be relatively stable" during this period (Australian Government 2008: xix–xx).

The initial CPRS called for a cap-and-trade scheme that would set a limit on total annual emissions and stipulated a carbon price of $10 per ton in the first year. It proposed to cover some 1,000 companies, as well as emissions from stationary energy, transport, industrial processes, waste, and "fugitive emissions," such as waste emissions resulting from oil and natural gas production. The initial CPRS stipulated that in "emissions-intensive trade exposed (EITE) industries, producers of goods for transport" would be granted 90 percent of their permits free, and other businesses would get 60 percent of their permits free. Despite many concessions to industry, various corporate groups, such as the Minerals Council of Australia and the Australian Industry Group, either opposed the scheme or wanted its introduction delayed. To win over the Coalition under then Opposition leader Malcolm Turnbull, the Rudd government announced a modified proposal for an ETS known as CPRS II on May 4, 2009. Apart from delaying implementation of the scheme by a year, which the Coalition had always favored, the new scheme proposed a 5 to 25 percent reduction band, a new $10 cap on carbon pricing, and more free trading permits for EITE industries. However, in large part because a substantial number of Coalition parliamentarians opposed even CPRS II, Tony Abbott managed to depose Turnbull as Opposition leader. Abbott repeatedly asserted that the CPRS constituted a "great big new tax." He offered an "environmental policy" rather than a "tax policy." Abbott's plan proposed to subsidize companies for voluntarily cutting greenhouse gas emissions and thus reducing 140 million tons of CO_2 by 2020. In achieving such a cut, the Coalition also proposed "direct action on soil carbons" and tree planting. The CPRS was defeated in the Senate on August 13, 2009.

In late 2010 and early 2011, much of Australia shifted from a land of drought and heat waves to one of cyclones, heavy rains, and floods. Indeed, worldwide 2010 turned out to be the wettest year on record but also a very hot year. The World Meteorological Organization maintained that 2010 was the hottest year since records began in 1850, and NASA and the National Ocean and Atmospheric Administration also reported that 2010 was the wettest year on record as well as a very hot year, tying with 2005 as the hottest year on record. Heavy rains and floods during late 2010 and early 2011 hit Queensland particularly and to a lesser degree the states of New South Wales and Victoria. Australia was ravaged by one of the most severe La Niñas in recorded history. David Jones, an Australian Government Bureau of Meteorology analyst, observed, "The last year of extreme weather events has been really extreme, but in the Australian context the really major story is La Nina" (Tippet and Russell 2011:9). La Niña in 2010 was aggravated by a record-high sea surface temperature, which very likely was related to global warming. At various levels, these events reminded Australians that they lived on a fragile continent.

The Rudd government ran into difficulties when the mining sector mounted a media campaign to counter a proposed Resource Super Profits Tax (RSPT). Julia Gillard, the former deputy prime minister, assumed the prime ministership in a set of circumstances that may have appeared to the rest of the world as a coup d'état. One of the first actions that she took as prime minster was to whittle down the RSPT from 40 percent to 22 percent and rename it the Minerals Resource Rent Tax—to be applied only to iron ore and coal companies that earn profits above $50 million (McAuley 2010:20). In arriving at this new arrangement, Gillard entered into negotiations with the three largest multinational corporations in the mining industry, BHP Billiton, Rio Tinto, and Xstrata. A Department of Treasury report indicated that the reformulated mining tax resulted in a forfeiture of $60 billion as compared with what the Rudd-proposed mining tax would have brought into government coffers (Thomas 2011:9). While Gillard reiterated Rudd's instance that climate change constitutes a serious threat to Australia and the world, she sent then minister of trade Simon Crean to Melbourne on June 25 to broker a deal to export brown coal to Vietnam (Butler 2010b:11).

Both the ALP and the Coalition downplayed climate politics during the 2010 election, which eventually permitted Gillard to continue on as prime minister in a minority government in which the Greens and the Independents came to hold the "balance of power." The Greens, who succeeding in having Adam Bandt from the seat of Melbourne elected as the first Green in the lower federal house, served as an important force in keeping climate politics alive during the 2010 election. Julia Gillard claimed initially that there would be no carbon tax under her leadership but later shifted her position to say a tax would transition into an emissions trading scheme (ETS). Finally, in March 2011 she admitted that she was speaking of a "carbon tax." The Multi-Party Climate Committee consisting of ALP and Green representatives proposed that the carbon price mechanism could begin as early as July 1, 2012, if it were to pass in both the House and Senate and could operate for three to five years, after which time it would convert into a "flexible cap-and-trade emissions trading scheme." Bob Brown maintained, "This agreement is the Greens in action, delivering certainty to the Australian economy, community, investors and the environment after productive negotiations with the government." Christine Milne viewed the plan as integral to the "transformation of our economy towards a zero-emissions future." In reality, the term *carbon price* as used in the Australian context conflates a carbon tax and an ETS/CPRS.

While the ALP and Greens joined forces in pushing for a carbon price mechanism, the Coalition accused the Gillard government of trying to impose "a great big new tax" on the Australian people. While Abbott had backed off for a while from his controversial statement that "climate change is crap," on March 23, 2011, he again exhibited his contrarian predilections when he told an audience in Perth that "whether carbon dioxide is quite the environmental villain that some people make it out to be is not yet proven" (quoted in Editors 2011b:3). Abbott spoke in front of a banner reading "Ju-Liar Bob Brown's Bitch" (an inference that Prime Minister Julia Gillard was carrying out the environmental agenda of Greens senator Bob Brown) at a rally protesting the carbon price mechanism in Canberra on March 2011. In the wake of an International Energy Agency report indicating that humanity had emitted some 30.6 billion tons of CO_2 in 2010, 5 percent more than the previous year, the Gillard government offered to boost it 2050 emissions reduction target from 60 percent to 80 percent in "exchange for compromises by the Greens on compensation for households, heavy industry and coal-fired power industry" (Arup and Morton 2011).

On July 10, 2011, dubbed Carbon Sunday, the Gillard government announced the details of its planned carbon price mechanism. It proposed the implementation of a carbon tax on July 1, 2012, which will transition into an ETS on July 1, 2015. The new Climate Change Authority to oversee the scheme will be headed by Bernie Fraser, former governor of the Reserve Bank. In May 2014 this body will set emissions caps for 2015 to 2019 and each year by June 30 will set a cap five years ahead. The cap, which limits the amount of carbon emissions, effectively sets the carbon price. It is designed to result in a 5 percent cut in Australia's CO_2 emissions, based on 2000 levels, by 2020 and an 80 percent cut in those levels by 2050. Some 500 big polluters will pay a tax to be set at $23 for every ton of CO_2 emissions, increasing to $24.15 in 2013 and 2014 and $25.40 in 2014 and 2015. Agricultural emissions are exempted. It is an extremely complex scheme that includes numerous subsidies for both industry and low- and middle-income households and free permits for some industries. The government estimates that the carbon tax will increase costs for the average household by 0.7 percent, or $9.90 a week, and result in a short-term 0.7 percent increase in inflation. The scheme compensates low-income households by lifting the tax-free threshold from $6,000 to $18,000 in 2012, then to $19,400 in 2015 and 2016, and grants a tax cut to all households below $80,000. The government projects that 2 out of 3 households will have their costs fully offset and that 9 out of 10 households will receive some compensation.

Compensation to industry was exceptionally generous. It allocates $9.2 billion over three years to support EITE industries; heavily polluting industries such as steel, aluminum, zinc, pulp, and paper will get "free permits" covering up to 94.5 percent of their emissions; $1.2 billion is allocated for a "clean technology program" to help manufacturers; $1.3 billion will assist affected coal miners in finding alternative employment. To encourage renewable energy, $10 billion is set aside to set up the Clean Energy Finance Corporation, which will invest in renewable energy projects, and $3.2 billion is allocated for research and development of renewable energy projects. The scheme intends for 20 percent of Australia's energy to come from renewable sources, such as wind, solar, and geothermal, by 2020, and large-scale renewable energy will provide 40 percent of electricity by 2050. The dirtiest generators will be paid to close, cutting 2,000 megawatts of coal-fired power by 2020. (Of a national total of $5.4 billion compensation to electricity generators, $5.22 billion, or 97 percent, is projected to go to generators in Victoria.) Gas-fired electricity will increase by more than 200 percent by 2050. Other funding initiatives include $429 million for a Carbon Farming Initiative for reforestation and revegetation schemes that would allow farmers to earn money by selling carbon credits to big polluters; $947 million over six years for a biodiversity fund to protect Australian native species from climate change; and $40 million over the next four years to promote energy-efficiency measures among small businesses and community groups.

Tony Abbott referred to it as a scheme for redistributing wealth. He reiterated Coalition policy that emissions should be cut by planting numerous trees, shifting from chemical to organic fertilizer, and improving energy efficiency, and he continued to support the coal-fired power industry. Legislation for the carbon price mechanism passed in the House of Representatives 74–72, assisted by votes from one Green member and four Independents, on October 12, 2011. Protesters in the public gallery yelled, "No mandate" and "Democracy is dead," and Abbott made "a pledge in blood" to repeal the carbon tax as the "first order of business" for an incoming Coalition government. On November 8, the Senate passed the carbon price mechanism, or "carbon tax," as it has become known in everyday parlance. Only time will tell whether this legislation will be a step forward—one that will move climate change mitigation forward in Australia, as the Greens, most environmental NGOs, and at least some climate activists assert—or merely create the appearance that effective action is being taken, thus circumventing more radical forms of action to create a safe climate not only for Australia but for the world.

The Emergence and Rise of the Australian Climate Movement

Climate action groups have grown exponentially since 2003 in response to the climate policies of both the Howard and Rudd governments and now constitute the most significant new social movement in Australian society, albeit one that has gone through ebbs and flows. Many of the CAGs are linked in complex ways to other organizations, such as established environmental nongovernment organizations (ENGOs), the Australian Greens (a growing political force in parliament), and extra-parliamentary political parties and groups, particularly the Socialist Alliance and Solidarity, another socialist group. Many of them have also become affiliated, beginning in early 2008, with two regional climate action networks and in early 2009 with a national climate action network.

The Australian climate movement is a dynamic but very disparate movement that consists of three broad layers, which can be classified respectively as acting "from above," "in the middle of," and "from below" society: (1) operating as a lobby group, largely within the corridors of power and concealed from public scrutiny, is the Southern Cross Climate Coalition (consisting of the Climate Institute, the Australian Conservation Foundation (ACF), the Australian Council of Trade Unions (ACTU), and the Australian Coalition of Social Services); (2) functioning primarily within civil society, though occasionally in parliaments and local councils, and within view of mainstream media are the Australian Greens, peak ENGOs such as Greenpeace, Friends of the Earth, the Wilderness Society, and Climate Action Network Australia (CANA), state-based nature conservation societies, and individual public intellectuals; and (3) acting at the grassroots level and below the radar of mainstream media are regional and local climate action groups and networks and socialist parties. Despite assertions of moving beyond a business-as-usual approach and acceptance of the climate science regarding the seriousness of climate change, including for Australia, the Rudd government in reality adopted a rather conservative approach on mitigation strategies, as has the successor Gillard government in an internal party coup of sorts. Some climate activists had become so angered by the Rudd government's failure to adopt adequate climate change mitigation strategies that they embarked on direct actions at a coal port, coal mine, and coal-fired power plants.

Actors in the Australian Climate Movement

Figure 8.2 identifies the layers and actors in the Australian climate movement. Indirectly, it illustrates the evolutionary process that social movements often go through if they achieve some level of institutional success. As Ariadne Vrommen, Katherine Gelber, and Anika Gauja (2009:267) observe, "Movement activity is believed to start as informal networking, then collective action such as protests occurs, then social movement organisations are established, then eventually there is a decline in protest activity and organisations consolidate to form long-term agenda." While some might argue as to which groups in the climate movement operate at the middle level of Australian climate politics and which operate at the grassroots level, it will become clear that some actors in the climate movement tend to be more bureaucratic and procedural, and some actors tend to be more spontaneous, egalitarian, and participatory. Nevertheless, even groups at the grassroots level of climate politics may exhibit tendencies toward bureaucratization and compromise with the powers that be.

Table 8.2. Layers and Actors in the Australian Climate Movement

Southern Cross Climate Coalition
- Climate Institute
- Australian Conservation Foundation
- Australian Council of Trade Unions
- Australian Coalition of Social Services

Mainstream Civil Society
- The Greens
- Peak Environmental NGOs
  - Greenpeace
  - Friends of the Earth
  - Climate Action Network Australia
  - World Wildlife Federation Australia
  - Wilderness Society
  - Safe Climate Australia
  - Oxfam
  - Brotherhood of St. Laurence
- State nature conservation councils

Grassroots Climate Movement
- Regional climate action networks
- Local climate action groups
- Special climate action groups
- Socialist groups
- Student environmental collectives

The Southern Cross Climate Coalition

The Southern Cross Climate Coalition (www.climateinstitute.org.au) is an alliance of the Climate Institute, the Australian Conservation Foundation, the Australian Council of Social Services, and the Australian Council of Trade Unions. Essentially it operates as a lobby group, largely within the corridors of power and concealed from public scrutiny. The Climate Institute (established 2005) defines itself as a "non-partisan, independent research organisation" committed to promoting "innovative and effective climate change solutions" (www.climateinstitute.org.au, accessed June 13, 2009). Historically, the Australian Conservation Foundation has been a moderate environmental organization, including on the issue of climate change. ACF called on the Australian government to set legally binding national targets to cut greenhouse pollution by at least 30 percent by 2020 (from 1990 levels). It has consistently supported the introduction of an emissions trading scheme that purportedly would drive down greenhouse gas emissions, avoid loopholes and giveaways, and generate revenues to invest in a transition to a "clean economy." ACF also advocate stabilizing greenhouse gases at 350 ppm CO_2e and cutting CO_2 emissions by at least percent 40 percent by 2020. It accepted the Rudd government's CPRS but withdrew its support on November 25, 2009, after the Rudd government agreed to compromise with Malcolm Turnbull by whittling it down. Conversely, the ACTU continued to support the CPRS despite these developments. For the most part, Australian trade unions have adopted a moderate stance on climate change mitigation. On the CPRS, the ACTU (2008:3)states,

> The union movement supports the commitment by the Federal Government to work to achieve a 60 per cent reduction in Australia's greenhouse gas emissions below the 2000 level by 2050. Our submission on the CPRS suggests a 30 per cent reduction in greenhouse gas emissions by 2020 is possible without major technological breakthroughs or lifestyle changes and at minimal cost to working people. The ACTU also supports the government's commitment to a medium term target for carbon emissions trading.

Like most mainstream NGOs and most labor unions, ACF supported the implementation of a carbon price mechanism.

Mainstream Civil Society

Various groups and organizations operating within the corridors of mainstream civil society, as well as occasionally in parliaments and local councils, and within the view of mainstream media, have come to address climate change. These include the Australian Greens, peak ENGOs such as Greenpeace, Friends of the Earth, the Wilderness Society, and Climate Action Network, and state-based nature conservation societies. Here I discuss the Greens and four ENGOs involved in the Australian climate movement: the Climate Action Network Australia, Friends of the Earth, Environment Victoria, and Safe Climate Australia.

The Greens

Particularly with the demise of the Australian Democrats in the Senate, the Greens have become more and more of a presence in parliamentary politics, in both the federal government and various state governments. The Australian Greens have generally made climate change one of their various environmental concerns. While not speaking officially on behalf of their party, Bob Brown and Peter Singer expressed concern about the "greenhouse effect" in the mid-1990s and proposed a carbon tax as a climate change mitigation strategy (Brown and Singer 1996). More recently, the Greens asserted that the Rudd government CPRS overlooked the risks associated with catastrophic climate change and essentially proposed to hand billions of dollar to the "big polluters" and undermine international efforts to avert global climatic and environmental disasters (Kaye n.d.). As part of the debate on the CPRS, the Greens submitted a Safe Climate Bill that called for a more stringent emissions trading scheme without free permits or compensation and with a higher carbon price. Components of the bill proposed "that, in the context of a global agreement, Australia's emissions will be reduced to at 40% below 1990 levels by 2020, and that the 350 ppm goal is unshrined in law." It stipulated provisions based on the following principles: "Making polluter pay, not paying polluters"; "repowering Australia with renewable energy"; "save energy, save money, save the planet"; "travelling with a light footprint"; and "protecting our green carbon" (www.safeclimatebill.org.au). On January 21, 2010, the Greens announced an "interim carbon price proposal" whereby CO_2 would be taxed within the framework of the CPRS. The Greens proposed taxing carbon at a rate of \$23 per ton in the first year, starting in July 2010, and \$24 in the second year. Green politicians, particularly Bob Brown and Christine Milne, have been speakers at climate change conferences and symposia in universities and climate action conferences and rallies, including the Climate Action Summits in Canberra. Both of them have become staunch proponents of the Gillard government's carbon price mechanism, as has Adam Bandt, a Green member of parliament from the seat of Melbourne.

Climate Action Network Australia

Climate Action Network Australia (www.cana.net.au) is a broad-based network that has numerous members, including Friends of the Earth, Greenpeace Australia, WWF-Australia, the Climate Institute, the Wilderness Society, the Australia Institute, the Australian Youth Climate Coalition, GetUp, Oxfam Australia, the Australian Religious Response to Climate Change, the Liquor, Hospitality and Miscellaneous Union, Environment Victoria, the Climate Emergency Network, Rising Tide, and several local CAGs. CANA had recommended 40 percent emissions cuts by 2020. Its policy paper on emissions reductions called for Australia's greenhouse gas emissions to peak by 2010 and drop by at least 40 percent by 2020 and by at least 95 percent by 2050 (with a baseline of 1990 levels). CANA criticized the Rudd government for low emissions targets, its willingness to allow the coal industry to continue functioning, and its advocacy of CO_2 capture and storage. It supported passage of the carbon price mechanism.

Friends of the Earth Australia

Verity Burgmann (2003:192–94) characterizes Friends of the Earth, an international NGO, as a "radical, egalitarian, unbureaucratic organisation, committed to sustainable activism on the ground" and the "most radical of the principal [Australian] green organisations." It " participated enthusiastically" in demonstrations against the World Economic Forum conference in Melbourne. Friends of the Earth Australia is a federation of approximately 10 local branches. It has maintained a vigorous Climate Justice Campaign based on the notion that "climate justice requires assisting the victims of climate change to adjust to their misfortune." Friends of the Earth has maintained a strong campaign to address the problems related to climate change faced in particular by the people of low-lying islands in the South Pacific. It has demanded "deep cuts," arguing that "Australia should set a national target of reducing emissions by at least 90–95% (from 1990 levels) by 2050." Friends of the Earth has called for a moratorium on biofuels, a carbon accountability alternative to carbon offsets, and changes in the transportation infrastructure. The organization expressed skepticism as to whether the adoption of an ETS "will be able to withstand political pressures of fossil fuel industry and other polluters who will attempt to reduce liabilities under such scheme." Conversely, Friends of the Earth admitted that an ETS could potentially serve as part of the solution to the growth of Australia's carbon emissions but only if it avoids the problems of previous schemes. Such an ETS would require that most sources of emissions be covered, as well as the exclusion of forest and soil sinks from the scheme, the auctioning of all permits, no granting of free permits to purported EITE industries, eschewal of the Clean Development Mechanism (CDM) and other offsetting schemes, and no borrowing of permits. Friends of the Earth also recommended stabilization of CO_2 concentrations at below 350 ppm (Friends of the Earth Australia 2008).

In responding to the proposed carbon price mechanism, Friends of the Earth Australia (2011) rejected the plan because of an anticipated low price on carbon and its belief that carbon trading "does not work—across the world cap-and-trade has failed to reduce emissions, harmed local communities and delayed real action; whilst delivering huge profits to polluters and financial speculators." Indeed, Friends of the Earth Australia has adopted an even more critical perspective on emissions trading schemes than it had adopted earlier. In its 2011 Climate Justice Position Paper, Friends of the Earth Australia (accessed March 21, 2011) adopted a sophisticated and nuanced position on various development theories. It argues,

> Conservative aid and development theory is based on a seemingly simplistic recipe for development known as "modernisation theory." Modernisation translates to that modern states

> [*sic*] that have a highly active economic sector based on a functioning capitalist market system are the ultimate form of development which all other states should strive for. . . .
> Dependency theory and World Systems theory has since identified that the "development" the third and fourth worlds has lead [*sic*] to over-exploitation of social and ecological resources for the consumptive benefit of the elite few in the South and the enriched communities of the first world.

Environment Victoria

Environment Victoria, along with the Nature Conservation Council of New South Wales, is a leading nature state conservation society involved in the climate movement. It published a report titled *Turning It Around: Climate Solutions for Victoria*, which demonstrates how Victoria can cut its greenhouse gas emissions 54 percent by 2020 (www.environmentvictoria.au/halving-our-emissions, accessed 01/18/2010). Environment Victoria identifies four areas in which emissions could be cut: (1) the promotion of sustainable products and consumption, (2) increased energy efficiency, (3) the creation of a low-carbon energy supply, and (4) an increased shift to public transportation. In a more recent report, Environment Victoria (2009) makes numerous recommendations to the Victoria government, including cutting emissions by at least 50 percent by 2020 from 1990 levels, switching from coal to natural gas as a "transition fuel," and creating new gas and renewable energy manufacturing jobs. It has been the principal organizer of the Walks against Warming in Victoria.

Safe Climate Australia

Al Gore launched Safe Climate Australia (www.safeclimateaustralia.org) at a Business Breakfast for a Safe Climate in July 2009 in Melbourne, which brought together some 1,000 "community leaders," including those representing "some of the largest and most influential corporations in Australia" (Safe Climate Australia n.d.). Safe Climate Australia defines itself as an NGO "formed and steered by a foundation group of concerned scientists, community and business leaders with a shared understanding of the scientific imperatives for emergency action to restore a safe climate" (Safe Climate Australia 2009). It acknowledges that climate change threatens the Australian environment and economy. Safe Climate Australia seeks to "identify and catalyse action on the societal transformations and solutions needed to achieve a safe climate" for both Australia and the planet. Its climate change mitigation strategies clearly fall very much within the parameters of ecological modernization, as is exemplified by its *Safe Climate Transition Plan,* which promotes as renewable and public transport. Safe Climate Australia seeks to develop a comprehensive transition plan to create a "zero-carbon economy." It supports a cut in CO_2 levels to between 280 and 325 ppm in order to create a "safe climate."

The Grassroots Climate Movement in Australia

The grassroots climate movement, which for the most part operates below the radar of mainstream media, includes numerous local CAGs, a still-emerging national climate action network, two regional climate action groups, various socialist groups, and student environmental collectives. The great majority of grassroots groups involved in the climate movement belong to a green social democratic wing that urges lobbying politicians and persuading business people to embrace a form of regulated green capitalism that would result in a reduction of greenhouse gas emissions. They favor a program of ecological modernization that would entail a strong shift to adopting renewable energy sources (particularly solar, wind, thermal, and wave power), energy efficiency, and mass transportation. While generally critical of the CPRS that the Rudd government tabled for discussion until 2013, most of the green social democratic groups are open to a stringent ETS scheme of one sort or other.

A much smaller democratic eco-socialist wing maintains that meaningful climate change mitigation ultimately requires the transcendence of global capitalism, which fosters a treadmill of production and consumption and economic expansion heavily reliant on fossil fuels. Actors in this wing include members of the Socialist Alliance, Solidarity, the Socialist Alternative, the Socialist Party Australia, and the Freedom Socialist Party, eco-socialists within the Green Party and perhaps the ALP, and an assortment of independent eco-socialists. Eco-socialists, however, recognize that the mitigation process cannot await the collapse of global capitalism, although some point to the Global Financial Crisis as a possible sign of the unraveling of this system. In the interim period, rather than simply awaiting the "revolution," the eco-socialists call for various transitional reforms, such as public ownership of utilities, the expansion of public mass transportation, minimization of car use, renewable energy sources, and redistributive mechanisms. Whereas the Socialist Alliance has been leaning toward the implementation of a steep carbon tax at the site of production, with rebates for low-income people, Solidarity vehemently opposes a carbon tax.

Local Climate Action Groups and the Drive for a National Community Climate Network

The origins of local climate action groups, such as Rising Tide in Newcastle, in Australia remain obscure. Figures on the number of climate action groups in Australia vary. The climate network website (www.climatemovement.org.au) lists over 100 CAGs, but others have claimed the existence of some 280 CAGs. As Figure 8.3 illustrates, the Australian grassroots climate movement has engaged in a wide array of activities, ranging from conferences to rallies and climate camps to lobbying politicians and business people.

The Climate Action Summits of 2009 and 2010 were part and parcel of an effort to create a national community climate action network. The Climate Action Summit held at the Australia National University on January 31 to February 2, 2009, was organized by a committee of activists from the Australian Student Environment Network, Rising Tide (Newcastle), Greenpeace,

Table 8.3. Activities of the Australian Climate Movement

Educating
- Meetings
- Conferences
  - Climate Justice Seminar at Melbourne University in August 2008
  - Climate Action Summit in Canberra, 2009 and 2010
  - Climate Action Summit in Melbourne, 2011
- Online information

Rallies and Climate Camps
- Climate Action Rally at Parliament House
- National Climate Emergency Rally in capital cities and other cities, June 2009

Climate Camps followed by Rallies
- Newcastle, Hazelwood, Port Augusta, Helensburg, Western Australia

Lobbying of Politicians and Business People

Friends of the Earth, and the Canberra Climate Action Network, along with other individuals. In part, the Climate Action Summit was designed to coordinate the numerous local climate action groups, many of which have focused on individual solutions, such as promoting solar panels.

The summit consisted of several plenary sessions and numerous concurrent workshops. Meals consisted of catered vegan food. Clive Hamilton, one of Australia's leading public intellectuals and a former director of the Australia Institute, a left-wing think tank, gave a speech titled "The Time for Radical Action" at the Climate Action Summit in its opening session on January 31, 2009. He observed, "There is an unholy alliance between government and industry to defer and delay action, to deny the true implications of global warming, and to hoodwink the public into sharing their view that protecting the old energy industries must come first. Except for the Greens, the main political parties have been captured by the fossil fuel lobbyists and climate skeptics" (Hamilton 2009:3).

Of 81 recommendations made at the summit, 72 received majority support. Nine recommendations, including ones on nuclear power, carbon capture and storage, and biofuels, were rejected. The summit passed a resolution demanding that the

Australian government participate in an international effort to reduce global levels of carbon dioxide to 300 ppm no later than 2050; that Australia's CO_2 emissions be reduced by at least 60% by 2020 and 90% by 2030 (from 1990 levels); and that the government enact a policy of 100 percent renewable energy by 2020, establish a moratorium on all new coal- and gas-fired power plants immediately, revolutionize energy efficiency, promote mass transportation, foster agricultural biological resilience, and create a moratorium on native forest logging.

Participants at the Climate Action Summit had different views on the merits or flaws of emissions trading schemes in general but agreed that the CPRS is a flawed proposal, with its free permits and compensation for big polluters and low emissions reduction targets. The summit called for the creation of "green jobs." It also called for four key mass actions during 2009: at the expected presentation of the CPRS to the federal parliament in March; in June, on World Environment Day; in September regarding an expected Arctic melting; and leading up to the UN Copenhagen Climate Change Conference in December. No consensus was achieved on the structure for a national climate network. Although the Green Party was not one of the sponsors, Christine Milne, a Green senator from Tasmania, spoke at the summit and supported the actions of its participants. Bob Brown, also a Green senator from Tasmania and the national spokesperson for the Green Party, addressed the Climate Action Summit's rally at Parliament House on February 3. The summit's organizers claimed there were some 2,500 attendees at the rally, but the *Canberra Times* reported on February 4 the presence of 1,250 attendees.

The Community Climate Network, a national organization, emerged following much organizing when over 100 CAGs endorsed the effort in late 2009. The network organized the second Climate Action Summit, which occurred on March 13 to 15 at the Australian National University. Plenary speakers included David Karoly (a prominent climate scientist based at the University of Melbourne), Ian Fry (chief climate change negotiator for Tuvalu), Christine Milne (deputy leader of the parliamentary Greens), Damien Lawson (director of the

Victorian Climate Action Centre), Donna Jackson (Australia Nuclear Free Alliance), Mark Ogge (Beyond Zero Emissions), Walden Bello (executive director of Focus on the Global South), George Woods (Climate Action Network Australia), Genevieve Kelly (National Tertiary Education Union), and Clive Spash (a former CSIRO ecological economist). The summit was structured around six campaign streams: 100 percent renewables, coal campaigning, vote climate/election campaigning, climate emergency, and green jobs. The 2010 Climate Action Summit only drew some 300 attendees and consisted of six campaign streams: national climate action network, 100 percent renewables, "quit coal," trade unions and green jobs, and the climate emergency mode. The 2010 summit retained the previous summit's commitments to no CPRS, 100 percent renewable energy by 2020, and CO_2 300 ppm. The summit urged CAGs to campaign for candidates who support its climate policies. According to Ewan Saunders (2010, 11),

> The big support for a carbon levy was a significant change from last year's summit, at which a large majority of participants still supported a "well-designed" emissions trading scheme as the best option to curb greenhouse gas emissions. This year there was next to no support for carbon trading.

The summit adopted World Environment Day on June 5, 2010, as a national day of climate action revolving around the themes of renewable energy sources, ceasing coal production, safe climate transition, and green jobs. Most of the sessions and workshops at the Climate Action Summit focused on various aspects of ecological modernization, such as 100 percent renewable energy sources and stopping coal; various other sessions and workshops focused on topics such as "critiques of carbon trading, technology, and the growth economy," "food choices and climate change," "trade unions and green jobs," "alternatives to the CPRS," the "faith scene," and "climate justice, ethics, and organising for the climate movement" (Climate Action Summit 2010 Program).

Climate Action Summit 2011 took place on April 9 to 10 at the University of Melbourne. Some 300 people representing over 100 climate action groups attended the conference to hear various climate activists, scholars, and climate change experts discuss and debate climate action strategies. Much of the discussion in the panels and in the halls revolved around the proposed carbon price mechanism and whether climate activists should support a defective scheme or oppose it because of its inadequacies. In a communiqué released to the media, the summit declared that "current climate policy options in Australia are inadequate and bear little relationship to actions demanded by the science of climate change." The Australian Climate Action Summit (2011) called on the government to "develop and implement a comprehensive national climate policy framework that includes a national plan for transitioning to a zero emissions economy, with substantial reductions in energy use"; "make polluters responsible by ceasing subsidies for fossil fuels use immediately, and not compensating polluting industries"; "redirect subsidies from animal agriculture to sustainable plant-based farming and to just transition programs for affected workers and communities"; "provide generous support providing a just transition for affected workers and communities via income redistribution and/or direct assistance or other measures"; and "place a much greater focus on so-called 'complementary' measures."

Regional Climate Action Networks

Climate action regional networks are a relatively recent development and may have started on February 9, 2008, at the Climate Movement Conference that convened at Northcote High School in Melbourne. While most speakers and workshop organizers at the conference proposed strategies of adaptation and mitigation that clearly sought to address climate change within the parameters of "green capitalism" and advocated writing letters to and lobbying politicians and business leaders, many conferees seemed to be committed to mass actions and moving beyond "business as usual." A highlight of the conference was the launching of a report titled *Climate Code Red: The Case for a Sustainability Emergency* authored by David Spratt (CarbonEquity) and Philip Sutton (Greenleap Strategic Institute) and sponsored by Friends of the Earth Australia (Spratt and Sutton 2008a). Spratt and Sutton proposed a framework for climate change campaigning based on the argument that the time to address climate change is now and urgently requires emergency measures like those adopted by the Allied powers against the Axis powers during World War II. The launch of the report led to the formation of the Climate Emergency Network, an organization based in Victoria. Spratt and Sutton (2008b) expanded their monograph into a book that achieved widespread attention within Australia but also outside Australia to some extent.

At the moment, regional climate emergency networks exist in only two states, Victoria and South Australia. Climate Emergency Network based in Victoria has sought with mixed success to serve as an organization linking local climate action groups. CEN provided financial support for the group Beyond Zero Emissions, which has been involved in designing a "near-zero emissions/10-year transition plan" for Victoria and eventually all of Australia. The group's work has focused on technological innovations in the stationary energy, transport, land-use, housing and commercial building, industrial processes, and export revenue sectors of the economy. CEN, in collaboration with Friends of the Earth Australia, launched in 2009 an ambitious Transition Decade campaign committed to creating a "safe climate outcome" that would reduce the Earth's temperature by between 0.3°C and 0.8°C and reduce the CO_2 in the atmosphere between 280 and 320 ppm.

The Climate Emergency Action Network (CLEAN; www.cleansa.org.au) largely originated out of a conference held in conjunction with the Australian Education Union titled "Climate Emergency—No More Business as Usual" in Adelaide in October 2008. Members of the Ecosocialist Network in Adelaide played a key role in the formation of CLEAN. While, like CEN, CLEAN emphasizes the adoption of renewable energy sources, energy efficiency, and other technological innovations, largely as a result of a strong eco-socialist and eco-anarchist presence in its membership, it touches more on social justice issues than does its Victorian sister organization. In its positions statement drafted on November 18, 2009, the following three items strongly addressed social justice issues:

- "As the wealthy nations of the world have very largely caused the climate emergency and have unfairly exploited many of the poorest nations' people and natural resources, the wealthy nations owe a very substantial duty to compensate materially poorer nations by providing them with financial aid to develop green energy infrastructure and to enable them to raise their standards of living to reasonable levels."
- "Wealthy nations, due to their privilege and economic power, must play a leading role in mitigating climate change and be willing to make much more substantial cuts in their greenhouse gas (GHG) emissions, in absolute and relative terms, than poorer nations."
- "In dealing with the climate emergency, all solutions to the emergency must be based on the principles of both ecological sustainability and social and economic justice between all nations and between individuals within nations" (CLEAN SA position statement at 30/05/11, www.cleansa.org.au/).

Like most grassroots climate action groups, CLEAN supports the goal of 100 percent renewable energy infrastructure by 2020 and opposes the CPRS and nuclear power plants. It did support the implementation of an "effective carbon pricing mechanism," particularly a carbon tax.

Socialist Parties and Groups

Various socialist parties or groups, in particular the Socialist Alliance and Solidarity, have become key players in the climate movement. Members of socialist groups are often visible at climate rallies and conferences and have been active particularly in the CEN in Victoria, CLEAN in South Australia, and the

Climate Action Summit Network. Socialists have also played a role in the efforts to create a national climate action network.

The Socialist Alliance is an outgrowth of the Democratic Socialist Party or later the Democratic Socialist Perspective. The Socialist Alliance in its newspaper *Green Left Weekly* regularly publishes articles on the impact of climate change on human societies, including Australia, and climate politics at both the international level and in Australia. For example, Kamala Emanuel (2009:3) proposes an eco-socialist alternative to the CPRS in which the "transition to a post-carbon economy would entail job creation, with many new jobs in research, training, renewable energy, public transport, and manufacturing" and "direct government investment in energy efficient public transport and public housing." The Socialist Alliance has developed a Climate Charter that calls for an emergency plan to mitigate climate change. Its 10-Point Climate Action Plan proposes the following measures:

1. Introducing emissions reduction targets of at least 5 percent per year and achieving "100% renewable energy by 2020"
2. Implementing an international agreement that aims to "get all countries to agree on a global target of at least 90% emissions cuts on 1990 levels by 2030"
3. Shifting to a "zero-waste economy"
4. Requiring that existing houses and commercial buildings as well as all new buildings become energy efficient
5. Implementing public ownership of utility industries, immediate phasing out of coal mining and coal-fired power plants, providing jobs and retraining to workers in affected industries, and operating the "maximum possible base-load power from existing natural gas and/or hydro power stations instead of coal until renewable energy is available"
6. Nationalizing the car industry and retooling it to "manufacture wind turbines, public transport vehicles and infrastructure, solar hot water, [and] solar photovoltaic cells"

7. Constructing solar thermal plants and wind farms and upgrading the "national grid to make it compatible with 100% renewable energy"
8. Banning old-growth forest logging and creating a program of "reforestation, carbon farming and biodiversity protection"
9. Phasing out industrial farming reliant on oil-resource fertilizers, pesticides, and fuel
10. Providing free public transport, upgrading existing public transport services, nationalizing and upgrading interstate and ferry services, and facilitating cycling

The Socialist Alliance released a statement on March 5, 2011, titled "A Carbon Price Does Not Equal Climate Action." The statement asserts that the "highest prices now being discussed will simply stimulate a mass rollout of gas, extending Australia's commitment to fossil fuels at the expense of renewable energy." It argues that while a "carbon tax can be a secondary or additional aspect of the government's climate policy," the "Socialist Alliance stands for large-scale public investment in a publicly-owned and run renewable energy sector, a rollout of public transport, a shift to sustainable farming and other carbon abatement programs, as well as government regulation to phase out dirty industries." It also maintains that the "the government needs to stop fiddling with the market to cut carbon emissions and instead set emissions cuts targets based on the science."

While not as large as Socialist Alliance, Solidarity has made climate change one of its principal campaign streams. Solidarity formed in February 2008 when three social groups—Solidarity, Socialist Action Group, and the International Socialist Organization—merged. Solidarity regularly publishes articles on climate change and climate politics in its monthly magazine *Solidarity*. In early 2010 Solidarity stated its opposition to the ALP/Green carbon price mechanism in part because the polluters will pass the costs incurred by a carbon tax on to consumers. It maintained that the "climate movement will not be able to win the mass support needed for serious action by telling workers to take a cut in their living standards" (Solidarity 2011a: 5).

Special Climate Groups

Particularly prominent among the special climate groups are Rising Tide and the Australian Youth Climate Coalition. On its website, Rising Tide Australia states that it is a

> grassroots Newcastle group taking action against the causes of anthropogenic climate change and for equitable, just, effective, and sustainable solutions to the crisis. We are committed to the principles of Non-violent Direct Action. We are part of the global Rising Tide climate justice movement. We live in the biggest coal port in the cosmos.

The Australian Youth Climate Coalition was created by lawyers Amanda McKenzie and Anna Rose in November 2006. It consists of some 27 youth organizations, including the Australian Medical Student Association, the Australian Student Environment Network, Engineers without Borders, National Indigenous Youth Movement of Australia, and the Sai Youth (Hindu Youth Network).

An Overview of the Climate Movement in Australia

The climate movement has been divided on whether virtually any type of ETS, such as the CPRS or the proposed carbon price mechanism, would be better than none at all. In 2011 an intense debate erupted within the climate movement over the proposed carbon price mechanism. While some climate activists accept the dilemmas of an ETS, they maintain that the movement should support the proposed legislation to get a carbon tax and then campaign against its rolling over into an ETS in three to five years. Whereas the Australian Council of Trade Unions and various environmental NGOs, such as the Australian Conservation Foundation, the Climate Action Network Australia, and the Brotherhood of St. Laurence, joined the Greens in supporting the carbon price mechanism, many grassroots climate activists, such as ones in the Socialist Alliance, oppose it (Betzien 2011:10). In contrast, while most participants at the Climate Action Summit that met at the University of Melbourne in April 2011 expressed qualified support for a "carbon price," they regarded the ALP/Green carbon price mechanism as grossly inadequate as a climate change mitigation policy; still, many attendees favored it as a way of moving forward on climate action.

Beyond Zero Emissions and other groups in the climate movement have expressed that Gillard is committed not to renewable sources of energy but to a shift from coal power to natural gas power. Indeed, on March 9, 2011, Greg Combet stated, "For baseload electricity generation it will be a gas-fired electricity that we see emerge, and for that investment to be committed, we need a carbon price in the economy" (Moore 2011:7). Indeed, the government has approved exploration for coal seam gas sites in Queensland, Western Australia, Victoria, South Australia, Tasmania, and the Northern Territory.

The disparate nature of the Australian climate movement, even at the grassroots levels, raises certain questions about the social scientific portrayal of new social movements (NSMs). For example, Keith Faulks (1999:88) maintains that the "novelty of NSMs can be seen in their disillusionment with the statist politics of the socialist left and the neo-liberal right, and their explicit rejection of the state as a tool that can be utilised to create social justice and ensure democratic accountability." In the case of the Australian climate movement, many climate activists adhere to the notion that they can persuade politicians and even corporate elites to reform the political economy in such a way as to prevent dangerous climate change. Furthermore, there are socialists involved in the movement who, although eschewing efforts to appeal to the corporate elites to act in a socially responsible manner on climate change, have engaged in electoral politics. Besides the socialist groups and some groups such as Friends of the Earth and a few others, a major weakness of the Australian climate movement is its muted critique of global capitalism and emphasis on social justice issues. This contrasts quite starkly with various other climate action groups around the world, particularly in developing countries, many of which might be more aptly termed *climate justice groups*.

Transforming the Climate Movement into a Climate Justice Movement

Christine Frank criticizes much of the environmental movement as being too soft on capitalism. She calls for an "uncompromising environmental movement led by working people in alliance with other oppressed groups," one that is imbued with "ecosocialist principles that go beyond the maintenance of capitalism and its suicidal and genocidal policies, and instead advance a zero-growth, zero-waste, steady-state, democratically planned socialist economy that puts planetary and human needs before profits" (Frank 2009:43). Wallis (2006:39) argues that the impact of the environmental movement has been small compared to the lifestyle changes needed to achieve environmental sustainability. He cautions that a "negotiating space" is needed between satisfaction of individual desires and needs and overarching guidelines (Wallis 2006:40). The creation of this space entails caution in issuing pronouncements about activities or technologies, such as long-distance travel, computers, and air-conditioning. Wallis (2006:41) cautions that a strident environmental moral stance could easily backfire.

Amory Starr (2003) has studied 15 transnational social movements that name corporate capitalism as the enemy. He divides these movements into three categories: (1) those that engage in "contestation and reform," such as the human rights and peace movements and cyberpunks; (2) those that promote "globalization from below" and "populist global governance"; and (3) those that seek to delink local communities from global capitalism, thereby building "small-scale communities that are protected from corporations." Chase-Dunn (2005:183) argues that what is needed is globalization from below, which "means the transnationalization of antisystemic movements and the active participation of popular movements in global politics and global citizenship." Chase-Dunn (2005:184) argues, "One of the big challenges is how the different kinds of progressive social movements can work together to struggle against capitalist globalization." He argues that "socialism or anarchism within one country or one community will not work for very long, and that we must confront the issues of global governance head on in order to move toward a more humane and equitable world society" (Chase-Dunn 2005:184). In his view, major transnational antisystemic movements consist of the labor, women's, environmental, and indigenous movements (Chase-Dunn 2005:185). Of these, the environmental and women's movements have been most successful in creating transnational linkages, whereas the labor and indigenous movements have thus far been less successful in doing so. Chase-Dunn (2005:187) maintains that a "truly democratic global peacekeeping government should be the eventual goal of the family of antisystemic movements."

The international climate justice movement per se has existed at least since the early 1990s. For example, at COP4 in Buenos Aires in 1998, the Climate Action Network, with affiliates in many countries, "drew media to its 'Fossil of the Day Award' and organized seminars on justice and equity with regard to emissions of greenhouse gases" (Pattberg and Stripple 2010:137). The climate justice movement appeared in full force in 2000 at the Climate Justice Summit that convened outside COP6 at The Hague, when it declared that "the causes of climate change are the production and consumption patterns in industrialised countries." The India Climate Justice Forum (2002:1) at the Climate Justice Summit on October 2002 in New Delhi stated the following:

> We, representatives of the poor and the marginalised of the world, representing fishworkers, farmers, Indigenous Peoples, Dalits, the poor and the youth, resolve to actively build a movement from . . . a human rights, social justice and labour perspective. We affirm that climate change is a human rights issue. . . . We reject the market based principles that guide the current negotiations to solve the climate crisis: Our World is Not for Sale!

The Durban Group for Climate Justice conducted a seminar in South Africa in October 2004 that was convened by the Dag Hammarskjöld Foundation in collaboration with various civil society organizations. Carbon Trade Watch is a project of the Transnational Institute (www.tni.org). Even more progressive religious groups have become part and parcel of the climate justice movement. At the UN Framework Convention on Climate Change (FCCC) conference in Montreal in December 2005, the World Council of Churches (2005) released a statement, "A Spiritual Declaration on Climate Change," asserting, "We commit ourselves to help reduce the threat of climate change through actions in our own lives, pressure on governments and industries, and standing in solidarity with those most affected by climate change."

Other organizations involved in the international climate justice movement include the World Rainforest Movement and the Greenhouse Development Rights Network directed by Tom Athanasiou and Paul Baer. Various indigenous groups have also become involved in the climate justice movement. One of these is the Inuit Circumpolar Council (ICC), which was established in 1977 in Barrow, Alaska, as a body that represents Inuit people from Greenland, Canada, and Alaska (Stern 2010:175). The ICC participated in the Earth Summit in Rio de Janeiro in 1992 and the 2002 World Summit on Sustainable Development in Johannesburg. In addition to protesting persistent organic pollutants, such as DDT, PCBs, and dioxins, that adversely impact Inuit people, the ICC has expressed strong concern about the impact of climate change on Inuit people. In 2004 Canadian Sheila Watt-Cloutier, then international ICC chairperson, testified before the US Senate Committee on Commerce, Science, and Transportation about the impact of climate change on Inuit communities, arguing that climate change is not only an environmental issue but also a human rights one and that the Inuit have a "right to be cold" (quoted in Stern 2010:183). In December 2005, she led a delegation of 62 Canadian and American Inuit in filing a petition to the Inter-American Commission on Human Rights, an affiliate of the Organization of American States (OAS), maintaining that the United States has violated Inuit rights by failing to reduce greenhouse gas emissions contributing to global warming. The OAS rejected the petition but asked Watt-Cloutier to testify at a March 2007 hearing. She was nominated for her efforts along with Al Gore and the IPCC for the Nobel Peace Prize in 2007, but unlike Gore and the IPCC, she was denied this distinction. The Venezuelan indigenous group Homo et Natura is part of a campaign resisting the TranGuajira 'Poliduct' operation that "supports big coal and big oil in the region, the Bolivar or America Harbor, and the rail lines that are part of the coal industry expansion in Zulia" (Cooke 2010:426). The Ecosocialist International Network had a meeting at the World Social Forum in Belém, Brazil, between January 27 and February 1, 2009, in which it drafted a Belém Ecosocialist Declaration, noting the need to make strenuous climate change mitigation efforts:

> To avoid global warming and other dangers threatening human and ecological survival, entire sectors of industry and agriculture must be suppressed, reduced, or restructured and others must be developed, while providing full employment for all. Such a radical transformation is impossible without collective control of the means of production and democratic planning of production and exchange. Democratic decisions on investment and technological development must replace control by capitalist enterprises, investors and banks, in order to serve the long-term horizon of society's and nature's common good. (Quoted in Wall 2010a:165)

Klimaforum, which met outside the Copenhagen Climate Conference in December 2009, and the World People's Conference on Climate Change, which met in April 2010 in Bolivia, have posited that the roots of climate change are embedded in global capitalism, which ultimately must be transcended. La Via Campesina, an organization with some 80 affiliates around the world, was part of the call for actions at the Copenhagen Climate Conference in December 2009 and argued that the UN FCCC "has failed to radically question the current models of consumption and production based on the illusion of continuous growth" (quoted in Tokar 2009b:7).

The presence of climate justice activists at COP15 in December 2009 was the result of more than 100 activists from 21 countries discussing such participation in September 2009 (Petermann and Langelle 2010:1). Additional organizational meetings occurred at meetings in Poznan, Poland (2008 UN Climate Conference); Belém, Brazil, during the 2009 World Social Forum;

and Copenhagen. Out of this process, Climate Justice Action (www.climate-justice-action.org) emerged as the principal organizing network for demonstrations at COP15. Various Danish organizations formulated People's Summit Klimaforum09, which features workshops, debates, discussions, and artwork on the need for an alternative world system. An estimated 50,000 to 100,000 people demonstrated on December 12, 2009, on the streets of Copenhagen on an International Day of Action. The Danish police used tear gas and batons in its efforts to subdue the demonstrators. Some 900 people reportedly were arrested (Petermann and Langelle 2010:3).

In the aftermath of the Copenhagen conference, Climate Justice Now! issued the following statement:

> The only discussions of real solutions in Copenhagen took place in social movements. Climate Justice Now!, Climate Justice Action and Klimaforum09 articulated many creative ideas and attempted to deliver those ideas to the UN Climate Change Conference through the Klimaforum09 People's Declaration and the Reclaim Power People's Assembly. . . . Copenhagen will be remembered as an historic event for global social movements. It will be remembered, along with Seattle and Cancun, as a critical moment when the diverse agendas of many social movements coalesced and became stronger, asking in one voice for system change, not climate change. (Quoted in Petermann and Langelle 2010:6)

Climate Justice Action called for a day of action for climate justice to take place on October 12, 2010.

Evo Morales, the president of Bolivia, convened the World's Peoples Conference on Climate Change and the Rights of Mother Earth in Cochabamba in April 2010. Over 35,000 people from 142 countries attended the conference at which Morales asserted that "either capitalism dies or Mother Earth dies." The People's Agreement drafted at the conference called on the developed countries to take the lead in returning the planet's greenhouse gases (carbon dioxide) to 300 ppm, thereby limiting the increase in the average global temperature to a maximum of 1°C. It further called for the creation of an International Climate and Environmental Justice Tribunal with the legal capacity to judge and penalize states, industries, and people with regard to their contribution, either through commission or omission, to climate change.

Climate justice activists have mounted numerous campaigns on a wide array of issues. For example, some campaigns have targeted the CDM, the voluntary carbon market, and various offsetting schemes, arguing that these practices constitute a form of *carbon colonialism*, or a "means by which rich consumers in the West merely displace their high-carbon consuming practices by buying offsets for their emissions cheaply the South" (Newell and Paterson 2010:32).

Zhou Shengxian, China's leading environmental official, reported some 51,000 pollution-related protests in 2005 in that country (Leonard 2008:43). In March 2007, Friends of Nature, Oxfam Hong Kong, Greenpeace, Action Aid China, Global Village Beijing, Worldwide Fund China, Green Earth Volunteers, and the Institute of Public Affairs launched the Chinese Civil Society's Response to Climate Change project (Wen 2010:140). In July 2007, 40 NGOs launched the 20 Percent Energy Saving Citizens Action initiative in keeping with the Chinese government's goal of improving energy efficiency 20 percent by 2020. Overall, the Chinese climate movement appears to be relatively weak in China. According to Alex Lo (2010:1015),

> Formal engagement in international climate networks is in an initial stage. Their first participation in the international climate conference in Bali in December 2007 led to a joint declaration with transnational NGOs, which was described as conservative and conforming to the Chinese government's position. Knowledge diffusion is the primary purpose of their involvement. Chinese NGOs hesitate to adopt the shaming strategies of western NGOs; the protest culture prevalent among international climate activists is not on their agenda.

Conclusion

Around the turn of the twenty-first century, a climate movement emerged in response to warnings of climate change emanating from the past two or three decades. Those involved included climate scientists, environmental groups, NGOs, indigenous peoples in the Arctic and South Pacific, and other Third World and faith-based communities. The international climate movement is quite disparate, as is the broader environmental movement that in large part it has evolved. Whereas mainstream NGOs and many climate action groups in developed countries have called for a program of ecological modernization, other environmental and climate action groups, as well as faith-based groups, both in developed and developing countries, are noting the need to consider social justice issues in addressing environmental degradation, including climate change. The global movement to create a safe climate and environmental sustainability remains in its infancy. The climate movement both internationally and within developed countries such as the United States and Australia needs to develop a climate justice analysis, one that recognizes the role of global capitalism with its strong emphasis on continual economic growth and its treadmill of production and consumption, both of which are heavily reliant on fossil fuels, as a generator of anthropogenic climate change.

As Mueller and Passadakis (2010:563) so succinctly assert in the last of their eight theses against green capitalism, "As an emerging climate justice movement, we must fight two enemies: on one hand climate change and 'fossilistic capitalism' that causes it, and on the other, an emergent green capitalism that won't stop it, but will limit *our* ability to do so." Daniel Tanuro (2009:263–66) calls for a multipronged movement to fight climate change, one that includes struggles for peace, women's rights, jobs, public provision of land, water, natural resources, the right of asylum, and indigenous rights, as well as against poverty, economic insecurity, privatization, and the globalization and liberalization of agricultural markets. To his list, one could add public provision of education and health care.

9

Conclusion

Historically, anthropologists have concerned themselves with human societies of the distant past, more specifically the domain of archaeology, and of the recent past or present, more specifically the domain of sociocultural or social anthropology or ethnology. In the 1970s various anthropologists began to call for an anthropology of the future, in part perhaps due to the fact that they adhered to the possibility of social liberation or emancipation. The demise of the Soviet bloc countries and the disillusionment with grand theory under the guise of postmodernism appear to have predisposed a younger generation of anthropologists to steer away from seemingly grandiose projects of attaining a better world both in terms of social justice and environmental sustainability. Yet a revival of the anthropology of the future, given an increasing awareness of the seriousness of anthropogenic climate change, strikes me as imperative. In contrast to many anthropologists who have opted to concern themselves with rather mundane debates, John Bodley (2008) has consistently, in the six editions of his *Anthropology and Contemporary Human Problems,* ended his seminal book with a chapter titled "The Future." He asserts that various social developments and movements are already pushing humanity toward a more egalitarian, democratic, and environmentally sustainable world system, one in which multinational corporations will gradually see their powerful dominance of world affairs diminish. In time, Bodley (2008) sees humanity shifting from macro structures, such as multinational corporations and large nation-states, to small nations that are more responsive to human needs, not only material but also social and emotional ones.

Obviously, eventually the human species, like so many other species that have inhabited this planet, will become extinct, if not due to developments of our own making then due to natural events over which we have no or little control. Yet it seems that we as a species can exert some degree of agency or control over our fate during the course of this pivotal century with respect to our some 5 to 6 million years on this planet. This will require a massive collective effort that will have to challenge and transcend a well-entrenched but nevertheless fragile capitalist world system. Although global capitalism has resulted in impressive technological innovations, it is a system fraught with contradictions, including an incessant drive for economic expansion; growing social disparities; authoritarian, militarist, and imperialist practices; depletion of natural resources; and environmental degradation (including global warming and associated climatic changes). It has become increasingly clear that human societies will have to adapt to the reality of climate change in a variety of ways, including technological innovations, reliance on renewable energy resources, significant expansion and improvement in mass transit systems, more efficient forms of heating and cooling, development of buildings and dwelling units that are more energy efficient, redesign of cities to control their energy demands and heat outputs, restoration of degraded environments, more sustainable agriculture, reforestation, protection of biodiversity, and less reliance on airplanes and private motor vehicles as forms of travel and less reliance on factory farming of animals. As important as these and other strategies will be in mitigating climate change, they are insufficient if not ultimately part and parcel of a longer-term effort to transcend global capitalism.

Perhaps more than any other issue, climate change allows critical social scientists to contemplate the contradictions of the existing capitalist world system and to contemplate the creation of an alternative world system, one committed to social equality, democracy, environmental sustainability, and a cooler planet, one in which humans can live in balance with one another as well as with animal and plant life.

In terms of climate, drawing upon the 2007 Intergovernmental Panel on Climate Change (IPCC) report, table 9.1 depicts selected examples of the projected impacts of climate change, scenarios that actually may be on the conservative side given the limitations of IPCC reporting (IPCC 2007b).

Table 9.1. Selected Regional Impacts Resulting from Climate Change

Africa

- By 2020, 75 to 250 million people will be exposed to increased water stress, and agricultural productivity may decrease up to 50 percent.
- By 2080, there will be an increase of 5 to 8 percent in arid and semiarid land in Africa.
- Sea level rise will severely impact large coastal cities, such as Lagos and Cairo.
- Climate change will lead to an increase in various infectious diseases, such as dengue fever, meningitis, cholera, and perhaps particularly malaria, with the incidence of malaria increasing in southern Africa and the East African highlands.

Arctic

- The Arctic ice cap will continue to diminish in both surface area and thickness, along with the thawing of the permafrost, which will result in devastating impacts on native communities as well as animal life over the course of the twenty-first century.
- Climate change will variably affect Arctic fisheries, with some seeing an increased yield and others a reduction.

Asia

- By 2050, there will be a significant decline in freshwater supplies in much of Central, South, East, and Southeast Asia.
- Various coastal areas, rural areas such as in Bangladesh, and coastal cities, such as Shanghai, Bangkok, and Manila, will face risks due to flooding from the sea and/or rivers.
- There will be an increase in diarrheal diseases and thus increases in morbidity and mortality in South and Southeast Asia.

Australasia

- By 2030, droughts may adversely impact southern and eastern Australia and northern and eastern New Zealand, resulting in greater probability of bushfires and decreased agricultural productivity.

Table 9.1. (*continued*)

- By 2050, coastal populations may be adversely affected by rising sea levels and increases in the number and severity of storms and flooding.
- As early as 2020, it is projected that there will be a significant loss of biodiversity in places such as the Kakadu wetlands of the Northern Territory and the tropical rainforest of Queensland and ongoing coral bleaching in the Great Barrier Reef.

Europe

- Most mountainous areas, such as the Alps, will see an ongoing pattern of glacial retreat.
- Southern Europe is projected to experience higher temperatures, more droughts, and a decline in agricultural productivity and tourism.

Latin America

- By 2050, increased temperatures are projected to result in desertification in the eastern Amazon Basin.
- The Andean glaciers are expected to continue to retreat, resulting in a shortage of water for human consumption, agriculture, and hydropower.
- Food security in many parts of the region is projected because of the decline of certain crops, although not necessarily soybeans, in temperate areas and animal production.

North America

- Temperature increases in western mountain ranges, such as the Rockies and Sierra Nevada, are projected to result in a diminishing snowpack, more frequent winter flooding, and reduced summer river flow.
- Various regions will be adversely impacted by an increase of pests, vector-borne diseases, and forest fires.
- Certain regions may experience an initial increase in agricultural productivity, but further warming could eventually result in a reversal of this pattern.
- Heat waves will increasingly plague large cities, such as New York, Chicago, and Los Angeles, as well as smaller cities and rural areas.
- Intense tropical storms will impact coastal communities.

As indicated earlier in this book, 2007 IPCC projections according to various other sources may be on the conservative side, on the matter of sea level rise. Researchers at the University of New South Wales Climate Change Research Centre in Sydney maintain the following:

> By 2100 global sea-level is likely to rise at least twice as much as projected by Working Group I of the IPCC AR4: for unmitigated emissions it may well exceed one meter. The upper limit has been estimated as [around] 2 meters sea level rise by 2100. Sea level will continue to rise for centuries after global temperatures have been stabilized, and several meters of sea level rise must be expected over the next few centuries. (Allison et al. 2009:7)

An increasing number of climate scientists assert that if drastic emissions cuts are not implemented soon, humanity may be headed for a world in which, by 2100 or sooner, the average global temperature is 4°C higher than it was at the time of the Industrial Revolution (Betts et al. 2011). In September 2009, a conference titled "Four Degrees and Beyond" took place at Oxford University in the United Kingdom. In the introduction to the conference proceedings, Mark New et al. (2011) state, "Even with strong political will, the chances of shifting the global energy system fast enough to avoid 2°C are slim. Trajectories that result in eventual temperature rise of 3°C or 4°C are much more likely, and the implications of these larger temperature changes require serious consideration." In July 2011, a conference titled "Four Degrees or More: Australia in a Hot World" (www.fourdegrees2011.com.au) was held at the University of Melbourne. In *Uncertain Futures*, Jonathan Ensor (2011:88) asserts,

> Under a 4°C scenario, uncertainty increases more rapidly with time, eroding confidence in future climate conditions more rapidly than in a world limited to 2°C warming. The prospect of successive incremental changes cascading into a process of near-continuous change means that even short lifetime decisions need to be informed by the context of climate change.

The Global Scenario Group was the brainchild of Paul Raskin, an American physicist who transformed himself into an energy analyst and scenario builder associated with the Stockholm Environment Institute. It is an independent, international, and indisciplinary body that has been focusing on global and regional scenario development, policy analysis, and public information dissemination. Drawing on earlier work of the group (Gallopin et al. 1997), Allen Hammond (1998:22–61) delineates three possible future scenarios: (1) Market World, (2) Fortress World, and (3) Transformed World. In the case of Market World, economic reform and technological innovation will stimulate rapid economic growth, which will supposedly result in "widespread prosperity, peace, and stability" (Hammond 1998:23). Fortress World results from the failings of Market World, which include social inequities and environmental disasters and results in a "future in which enclaves of wealth and prosperity coexist with widening misery and growing desperation, a future of violence, conflict and instability" (Hammond 1998:23–24). Finally, Transformed World results from policies and behavioral changes that humanize market relations and culminate in a global society in "which power is widely shared and in which new coalitions work from the grass roots up to shape what institutions [including presumably multinational corporations] and governments do," resulting in a more socially just and environmentally sustainable world (Hammond 1998:24).

In its later work, the Global Scenario Group refined its scheme to include three possible future scenarios for humanity with respect to the crisis of ecological sustainability, with two subscenarios in each of the broader scenarios, which are depicted in table 9.2 (Raskin et al. 2002).

Table 9.2. Possible Future Scenarios According to the Global Scenario Group

Conventional Worlds
Market forces: Adam Smith
Policy reform: John Maynard Keynes and commitment to "sustainable development"
Barbarization
Breakdown: Thomas Malthus
Fortress world: Thomas Hobbes
Great Transitions
Eco-communalism: William Morris, Gandhi, and E. F. Schumacher
New sustainability paradigm: John Stuart Mill

The Conventional Worlds scenarios entail two futures, one in which neoliberalism, with a faith in market forces, continues to serve as the prevailing wisdom and another that recognizes the need for some policy reforms under which governments and international bodies regulate the market to some degree in order to reduce poverty and ensure "sustainable development," as first delineated by the Brundtland Commission's report in 1987. Barbarization consists of two interrelated subscenarios in which conflict and crises lead to breakdown, resulting in the collapse of social institutions, a situation that becomes resolved through the creation of a Fortress World consisting of authoritarian institutions under which the "world divides into a kind of global apartheid with the elite in interconnected, protected enclaves and an impoverished majority outside," as exemplified by present-day gated communities (Raskin et al. 2002:15).

The Global Scenario Group report quickly glosses over the eco-communalism scenario, describing it as a "vision of bioregionalism, localism, face-to-face democracy and economic autarky" that is difficult to envision without global society first undergoing barbarization (Raskin et al. 2002:15). Instead the report views the New Sustainability Paradigm as the more attainable Great Transition scenario under which corporate globalization becomes humanized, resulting in a commitment to greater social parity and environmental sustainability (including heavy reliance on solar energy), the disappearance of advertising, the transformation of the United Nations into an authentic global federation, and the implementation of global electronic voting. Under this subscenario, corporations act in a socially responsible manner and are not strictly driven by profit making and a commitment to ceaseless economic growth. People will reside in cohesive communities in which they are situated relatively close to work, shopping centers, and recreational facilities. The Global Scenario Group believes that multinational corporations, the United Nations, the World Bank, the International Monetary Fund, and a wide array of NGOs will play key roles in achieving the New Sustainability Paradigm.

John Bellamy Foster (2005) laments that the Global Scenario Group fails to discuss its eco-communalism scenario in greater detail, which obviously would require an ecological revolution paving the way to something akin to democratic eco-socialism. Instead the focus of the Global Scenario Group on a New Sustainability Paradigm constitutes a "vision of the future that is contradictory to an extreme" (Foster 2005:9). As Foster (2005:9) observes,

> Private corporations are institutions with one and only one purpose: the pursuit of profit. The idea of turning them to entirely different and opposing social ends is reminiscent of the long-abandoned notions of the "soulful corporation" that emerged for a short time in the 1950s and then vanished in the harsh light of reality. Many changes associated with the New Sustainability Paradigm would require a class revolution to bring about.

Ecologist Peter F. Sale (2011) has recently delineated four possible future scenarios for humanity, which he calls Belvedere, Woodstock, Technopolis, and New Atlantis. Belvedere resembles the Fortress World scenario in that the privileged classes reside in "small, defended communities across North America, Europe, and Asia that will function as separate feudal states" and resemble present-day developed countries that "are already quite good at harvesting and sequestering environmental goods and services for their own benefit" (Sale 2011:284). Under the Woodstock scenario, humanity "solve[s] the problem of climate change by sustainably reducing [its] use of energy and returning to simpler lifestyles" (Sale 2011:284). However, Sale believes that the transition to a Woodstock-type world, with its emphasis on great social equality, is improbable unless key inspirational leaders emerge. In reality, the world is filled with inspirational thinkers, but unfortunately they generally find it difficult to gain entrée into existing governing institutions or are quickly co-opted if they do so. The Technopolis scenario that Sale (2011:287) envisions essentially relies on ecological modernization but may "become culturally sterile as increasing numbers of people live increasingly constrained lives." Recognizing that it would be difficult to achieve, Sale (2011:287) favors the New Atlantis scenario in which humanity decides to reduce its present population in order to enjoy "sustainably high standards of living across the world." New Atlantis would incorporate the best features of both Technopolis and Woodstock. In my view, achieving the zero-population growth or population stabilization of New Atlantis would require the eradication of poverty and the achievement of a highly egalitarian world system that would be part and parcel of democratic eco-socialism.

Obviously, counteracting the dystopian scenarios that climate scientists tell us are inevitable if humanity does not take drastic action on climate change will require major global and societal transformations. Over the course of the past seven years, during which I have concerned myself intensely with developing a critical anthropology of climate change, including one that looks at the impact of climate change on health (Baer and Singer 2009) and Australian climate politics, I have repeatedly found myself daunted by the immensity of the problem, which cannot be separated from the political economy of capitalism and its contradictions, including the social disparities and environmental degradation that stem from it. When people ask me to give them an example of a workable socialist society, I repeatedly note that socialism remains very much a vision and that postrevolutionary societies that have existed or continue to exist must be regarded as transitions between capitalism and socialism at best. While I would characterize Cuba in this manner, this tiny country some 90 miles from the United States constitutes a partial success story in terms of having achieved some semblance of social equality and environmental sustainability. Going from the present capitalist world system to an alternative global economy, whether it is defined as global democracy, economic democracy, Earth democracy, or democratic eco-socialism, will require much effort, and there are no guarantees that we will be able to create a more socially equitable and environmentally sustainable world. Conversely, do we really have any other meaningful choice, other than to continue on an ongoing downward spiral ending in the destruction of much of humanity and further environmental degradation, including climate change? Ultimately, mitigating climate change will require no less than what Foster (2009b:9) terms an *ecological revolution*—one that draws on the "struggles of working populations and communities at the bottom of the global capitalist hierarchy."

Beyond Kyoto: Toward a Progressive Global Climate Governance Process

Obviously any effort to create a global climate governance process will ultimately have to come from below. The World's Peoples Conference on Climate Change and the Rights of Mother Earth convened by Evo Morales in Bolivia in 2010 constituted a step in the right direction. Ultimately, the climate justice movement will have to form strong alliances with other progressive social movements, perhaps in particular the anti–corporate globalization or global justice movement. Immanuel Wallerstein (2011) contends that the capitalist world system has been in a state of chaotic crisis at least since the 1970s and probably will continue to be in one until around 2050. He asserts that this chaotic structural crisis "includes not only the world-economy, the interstate system, and cultural-ideological currents, but also the availability of life resources, *climatic conditions*, and pandemics" (Wallerstein 2011a:35; emphasis mine). In the competition for a successor system to the present one, he sees a struggle between two decentralized, disparate groups—namely, the "spirit of Davos" (World Economic Forum) group that desires some sort of "noncapitalist" system that is still hierarchical, exploitative, and polarized and the "spirit of Porto Alegre" (World Social Forum) group. The proponents of the spirit of Davos group are "divided between those who proffer the iron fist, seeking to crush opponents at all levels, and those who wish to co-opt the proponents of transformation by fake signs of progress (such as 'green capitalism' or 'poverty reduction')" (Wallerstein 2011a:37). The proponents of the spirit of Porto Alegre group are divided into "those who want a strategy and a reconstructed world that is horizontal and decentralized in its organization and insist on the rights of groups as well as individuals as a permanent feature of a future world-system" and "those who are seeking once again to create a new international [system] that is vertical in its structure and homogenizing in its long-term objectives" (Wallerstein 2011a:37).

The parties who acknowledge at some level that climate change is an anthropogenic problem can be placed into the four categories that Wallerstein delineates. Within the spirit of Davos group, one finds those who view climate change as a security risk for the core or developed countries and seek eco-authoritarian solutions and those who seek climate change mitigation and adaptation within governance structures such as the Framework Convention on Climate Change (FCCC) and the European Union and adhere to notions of green capitalism, sustainable development, and ecological modernization. Climate activists within the spirit of Porto Alegre group are not so easily categorized but tend to consist of two loose camps: those who espouse a clearly anticapitalist stance, either from an eco-socialist or eco-anarchist perspective, and those who are concerned about social justice issues but are not quite ready to discard capitalism.

In concluding this book, I find that, more than any other intellectual pursuit, my venture into the critical social science or anthropology of climate change has forced me to think outside the box while at the same wondering whether I am merely whistling into the wind or creating sandcastles in the sky. As Wallerstein (2011a:39) so eloquently states, "The one encouraging feature about systemic crisis is the degree to which it increases the viability of agency, of what we call 'free will.'" Hardt and Negri (2009:94–95) argue that "only movements from below" possess the "capacity to construct a consciousness of renewal and transformation"—one that "emerges from the working classes and multitudes that autonomously and creatively propose antimodern and anticapitalist hopes and dreams." The Arab Spring in the Middle East and the Occupy movement that emerged initially in developed countries but spread to developing countries in 2011 illustrate the power of agency in the face of seemingly insurmountable barriers. In that corporations, most governments, and transnational governance bodies such as the FCCC and the European Union have not been acting in a responsible manner in terms of serious climate change mitigation, despite much rhetoric to the contrary, much of the collective effort both to mitigate and to adapt to climate change will have to be spurred by progressive as well as explicitly antisystemic movements, including the climate justice movement.

David Harvey (2011:252) delineates "various broad fractious currents of thought on the left as to how to address the problems that now confront us." His listing includes (1) radical socialist and anarchist sects, which tend all too often to engage in fractious debates that mitigate against any meaningful effectiveness; (2) numerous NGOs that have appeared since the mid-1970s, which despite the presence of idealistic organizers generally refrain from adopting an explicitly anticapitalist stance; (3) anarchist and autonomist groups that often take an anticapitalist stance but tend to be ineffective in addressing global problems; (4) the more progressive strands of the labor movement; (5) groups based on the "need to resist displacement and dispossession (through gentrification, industrial development, dam construction, water privatization, the dismantling of social services and public educational opportunities, or whatever)"; and (6) identity movements, which focus on issues such as gender, sexual preference, ethnicity, and religion (Harvey 2011:256–57). Harvey (2011:278) asserts that an alternative to global capitalism is imperative, thus making the creation of a "global co-revolutionary movement" "critical not only to stemming the tide of self-destructive capitalistic behaviours (which in itself would be a significant achievement) but also to our reorganising ourselves and beginning to build new collective organisational forms, knowledge banks and mental conceptions, new technologies and systems of production and consumption, all the while experimenting with new institutional arrangements, new forms of social and natural relations, and with the redesign of an increasingly urbanised daily life." While Susanne Moser (2009:283) espouses more of a reformist than a revolutionary agenda, she argues "civil society can play at least two critical roles in climate change governance: (1) it can mobilize to push for policy changes at any level of government, and (2) it may enact behavioural changes consistent with needed mitigation and adaptation strategies." Ultimately the climate justice movement needs to join forces with other progressive or antisystemic movements that work together to create a postdemocratic global governance and that move beyond conventional representational governance structures in that they exhibit "flexibility and fluidity constantly to adapt to changing circumstances" (Hardt and Negri 2009:772). As anyone who has worked within social movements knows, despite their rhetorical espousal of democracy, in that their members have been socialized in hierarchical societies, there is often a tendency for certain individuals to attempt to dominate outcomes. I certainly have witnessed this firsthand in my observation of and participation in the Australian climate movement on more than one occasion. Ultimately, creating a more inclusive global governance process committed to social equality, environmental sustainability, and a safe climate will be a very complicated process for which no one individual has the answers but that will entail considerable debate and mutual respect of people engaged in the process.

I believe that the climate movement—or the climate justice movement, given that the former tends to downplay social justice issues—has a crucial role to play in such a global corevolutionary movement, but I also believe that critical anthropologists and social scientists have a part to play in this endeavor. In her list of issues that engaged anthropology examines, Kay Warren (2006:213) includes "social justice, inequality, subaltern challenges to the status quo, globalization's impacts, and ethnical positioning of our field research in situations of violent conflict." Ultimately, many of these issues are related to anthropogenic climate change, a topic that fortunately has become one of increasing anthropological inquiry (Crate and Nuttall 2009). In addressing this topic, it is imperative that anthropologists join other social scientists, such as sociologists, political scientists, and human geographers, in seeking to address the following questions in an act of collective responsibility that can be part and parcel of contributing to climate change mitigation, a safe climate, and climate justice:

1. The role of global capitalism in contributing to climate change

2. A vision of an alternative world system based on meeting basic social needs, social justice, and environmental sustainability, an effort that can serve as part and parcel of a critical anthropology
3. Identifying transitional reforms that can contribute to deeper systemic changes and social movements that are a part of this effort

It is my hope that this book has contributed in some modest way to achieving these goals.

Resource Guide

Research Centers

Hadley Centre (UK), http://www.metoffice.gov.uk/climate-change/resources/hadley

NASA Goddard Institute for Space Studies (USA), www.giss.nasa.gov

NOAA Climate Dynamics and Prediction Group (USA), www.dfdl.noaa.gov/research/climate

Pew Center for Climate and Energy Solutions (USA), www.pewclimate.org

Potsdam Institute for Climate Impact Research (Germany), www.pik-potsdam.de

Blogs

Climate Science Watch, www.climatesciencewatch.org
Real Climate, www.realclimate.org

Films

Changing Climates, Changing Times (Capa Drama, France, 2008). Directed by Marion Milne and Jean-Christoph de Reviere, this film focuses on four scenarios in which various protagonists in Europe, Africa, and Canada grapple with the impact of climate change on their lives in 2075. One story focuses on Idri and Faouzi and their incredible journey across the Sahara Desert to reach the Mediterranean Sea and make their way to Europe. Another story traces Grace, who anticipated the consequences of climate change in Arctic Canada but whose warnings were ignored. Another story focuses on Julia and her elderly father, whose vineyards are devastated by an unrelenting drought and heat in Bordeaux. In the final story, Lotte and her husband, Niels, seek to mobilize the world to make monumental changes in order to begin to reverse the damage inflicted on much of humanity by anthropogenic climate change.

Crude: The Incredible Journey of Oil (Australian Broadcasting Corporation TV, 2007). This 90-minute documentary examines the story of crude oil over the course of the past 160 million years. It examines the dependence of modern societies on oil in a multiplicity of ways, such as fuel for cars, industrial agriculture, and plastic products, in 11 countries and on 5 continents. It juxtaposes the Jurassic era—when dinosaurs roamed the Earth, oil first developed, the atmosphere contained several times the level of carbon dioxide that it does now, and the poles were free of ice—with the present era, in which human activities are producing greenhouse gases that may head the planet toward the atmospheric conditions of a long bygone era.

Gas Hole: A Crude Conspiracy (Time Life, United States, 2008). This film investigates the development of the oil industry and its impact on global society, as well as various possible solutions to oil dependence.

An Inconvenient Truth (Paramount, USA 2006). Narrated by Al Gore, this film won the 2007 Academy Award for the Best Documentary.

The Day after Tomorrow (Centropolis Entertainment, 2004). Loosely based on a scenario related to the theory of abrupt climate change in which global warming has caused massive shifts in the Gulf Stream, part of the Atlantic thermohaline circulation, which has halted. As a result, the North Atlantic region undergoes a drastic cooling while the tropics become incredibly hot. Large portions of formerly temperate areas experience a severe ice storm that suddenly plunges them into an ice age. In one scenario, a group of Americans make their way to Mexico to escape the catastrophe.

The Great Warming (Stonehaven Productions, Canada, 2006). Narrated by Keanu Reeves and Alanis Morissette, this film includes interviews with climate scientists and discusses the impact of climate change on people around the world, as well as various climate change mitigation strategies.

The Truth about Climate Change: What Is the Future of Our World? (BBC/Discovery Channel/Open University Coproduction, United Kingdom 2008). David Attenborough examines the debate as to whether recent climate change is a natural event, like climatic changes in the distant past, or is caused by human-related activities, particularly since the Industrial Revolution. The film depicts recent climate change–related events, such as Hurricane Katrina, the European heat wave of 2003, polar bears being forced to swim extraordinary distances from ice floe to ice floe and often drowning in the process, and huge swarms of insects descending on an African village. Attenborough believes that humanity must act quickly to prevent a global catastrophe.

Waterworld (Universal Studios, USA 1995). Fictional narrative about the distant future when climate change has melted the polar ice caps, forcing humans to adapt to a world almost completed covered with water. Kevin Costner plays the main protagonist, who protects a beautiful woman and a young girl from the dangers of pirates, particularly Deacon, who commands a rusted ship filled with hungry and desperate men in search of Dryland.

Novels and Fictional Accounts

Michael Barnes, *Mother of Storms* (1995). This science-fictional account is set in 2028 when nuclear warheads plunge into the

Pacific Ocean, resulting in the destabilization of huge amounts of methane clathrates, which in turn leads to global hurricanes. A highly ingenious astronaut attempts to save humanity by means of a geo-engineering scheme that will block sunlight.

Michael Crichton, *State of Fear* (2004). This gripping novel has served to reinforce the climate denialist cause and attacks the climate change movement for allegedly using a language of crisis to create unnecessary collective fear.

Kim Stanley Robinson, *Forty Signs of Rain* (2004), *Fifty Degrees Below* (2005), and *Sixty Days and Counting* (2007). This science-fictional trilogy discusses how the collapse of the North Atlantic conveyor belt due to climate change causes extremely cold weather in various areas and impacts on various characters, including in a snowbound Washington, DC.

Bruce Sterling, *Heavy Weather* (1995). This novel chronicles the pursuit of climate change–induced raging tornadoes across the US Plains by a group of technologically sophisticated storm chasers.

George Turner, *The Sea and the Summer* (1987). This gripping novel provides a dystopian portrayal of Australia, particularly Melbourne, in the 2040s as a result of a global temperature rise of 4.5°C above 1990 levels. Shortages of all sorts have become a normal part of Australian social life, but a class divide persists in Melbourne between the Swells and the Swills, with the former having retreated to the higher elevations and many of the latter residing in high-rise towers in the outer western suburb of Newport and inner eastern suburb of Richmond. Australia has, as a nation-state, relinquished its upper third to accommodate climate refugees from Asia. The Australian state has responded to the social chaos by becoming a totalitarian system with police intelligence officers seeking to keep the Swills under control in the various high-rise apartments.

W. Warren Wagar, *The Short of History of the Future* (1992). This fictional account depicts a dystopian world that begins to collapse with a nuclear holocaust in 2044, followed by two contrasting, partially utopian scenarios that emerge in the twenty-first and twenty-second centuries. Despite various technological innovations, which give the global economy brief spurts of growth, the capitalist treadmill of production and consumption continues to emit greenhouse gases contributing to ongoing climate change. Thus, by 2040, the atmosphere contains 555 ppm of carbon dioxide and even more alarming increases in methane, plus chlorofluorocarbons resulting from the burning of fossil and biomass fuels, fertilizer use, and the decay of organic matter in rice paddies. The average global temperature increases 7.56°F (4.2°C) between 1980 and 2040, resulting in ongoing melting of the polar ice caps and glaciers, a rise in sea level, heavy flooding in some regions, and the decline of food production. In the aftermath of the nuclear holocaust, the power center of the world falls to countries south of the twenty-fifth parallel; the World Party, with its technocratic socialist agenda, comes to power in much of the world and forms the Commonwealth in 2062, with Melbourne as its capital. Despite significant achievements, both in terms of social parity and environmental sustainability, the World Party gradually begins to lose credibility. In its stead, global society by 2157 consists of 41,525 autonomous communities of varying size, each with its own distinctive governance, economy, and social structure.

Bibliography

Abbate, Cinzia. 2008. "A Vicious Circle." In *Urban Climate Change Crossroads*, edited by Richard Plunz and Maria Paola Sutto, 128–34. Surrey, UK: Ashgate.

Agyeman, Julian. 2005. *Sustainable Communities and the Challenge of Environmental Justice*. New York: New York University Press.

Agyeman, Julian, Harriet Bulkeley, and Aditya Nochur. 2005. "Climate Justice." In *Ignition: What You Can Do to Fight Global Warming and Spark a Movement*, edited by Jonathan Isham and Sissel Waage, 135–44. Washington, DC: Island Press.

Albert, Michael. 2008. "Participatory Balkans: Exchanges between Andrej Grubacic and Michael Albert." In *Real Utopia: Participatory Society for the 21st Century*, edited by Chris Spannos, 136–44. Oakland, CA: AK Press.

Albert, Michael, and Robin Hahnel. 1991. *The Political Economy of Participatory Economics*. Princeton, NJ: Princeton University Press.

Ali, Syed. 2010. *Dubai: The Gilded Cage*. New Haven, CT: Yale University Press.

Alier, Joan Martinez. 2009. "Social Sustainable Economic De-Growth." *Development and Change* 40:1099–119.

Allison, Ian, et al. 2009. *The Copenhagen Diagnosis: Updating the World on the Latest Climate Science*. Climate Change Research Centre, University of New South Wales, Sydney, Australia.

Altavater, Elmar. 2006. "The Social and Natural Environment of Fossil Capitalism." In *Socialist Register 2007—Coming to Terms with Nature*, edited by Leo Panith and Colin Leys, 37–59. London: Merlin Press.

Amin, Samir. 2008. *The World We Wish to See: Revolutionary Objectives in the Twenty-First Century*. New York: Monthly Review Press.

———. 2009. "Capitalism and the Ecological Footprint." *Monthly Review* (November): 19–22.

Anderson, James. 2006. "Afterword: Only Sustain . . . the Environment, 'Anti-Globalization,' and the Runaway Bicycle." In *Nature's Revenge: Reclaiming Sustainability in an Age of Corporate Globalization*, edited by Josee Johnston, Michael Gismondi, and James Goodman, 245–73. Toronto: Broadview.

Anderson, Sarah, and John Cavanaugh. 2005. *Field Guide to the Global Economy*. New York: New Press.

Angus, Ian. 2010. "Ecocide and Consumption: A Case Study of Destructive Wealth." *Green Left Weekly*, August 18, 13.

Angus, Ian, and Simon Butler. 2011. *Too Many People? Population, Immigration and the Environmental Crisis.* Chicago: Haymarket Books.

Aronowitz, Stanley. 1994. "The Situation of the Left in the United States." *Socialist Review* 23(3): 5–79.

Arup, Tom, and Adam Morton. 2011. "Emissions Surge to Record Level: Labor Pushes New Plan on Carbon Tax." *The Age*, May 31.

Athanasiou, Tom, and Paul Baer. 2002. *Dead Heat: Global Justice and Global Warming*. New York: Seven Stories.

Australian Academy of Sciences. 2010. "The Science of Climate Change: Questions and Answers." Australian Academy of Sciences. August 2010. http://www.science.org.au/reports/climatechange2010.pdf.

Australian Bureau of Statistics. 2006. "Household Wealth and Wealth Distribution, Australia, 2005–2006." Australian Bureau of Statistics. http://www.abs.gov.au/AUSSTATS/abs@.nsf/Lookup/6554.0Explanatory%20Notes12005-06?OpenDocument.

Australian Climate Action Summit. 2011. "Not Good Enough!" Climate Action Summit 2012. April. http://climatesummit.org.au/communique.

Australian Council of Trade Unions. 2008. Principles and Policy on Global Warming. ACTU Position Paper, March.

Australian Government. 2008. "Carbon Reduction Scheme: Australia's Low Production Future." Canberra, Australian Capital Territory.

Australian Government Bureau of Meteorology. 2010. Annual Climate Summary 2010. Bureau of Meteorology. http://www.bom.gov.au/climate/annual_sum/2010/AnClimSum10_LR1.0.pdf.

Baer, Hans A. 2007. "Global Warming, Human Society, and Critical Anthropology: Toward a Research Agenda." Working Papers in Development Studies 1. University of Melbourne, Australia.

———. 2008. "Towards a Critical Anthropology of Global Warming: A Challenge for the 21st Century." In *The Impact of Global Warming on the Environment and Human Societies: Proceedings of a Symposium*, edited by Hans A. Baer, 15–38. PASI Research Papers. Melbourne, Australia: University of Melbourne.

———. 2009a. "A Field Report on the Critical Anthropology of Global Warming: A View from a Transplanted American Down Under." *Dialectical Anthropology* 33:79–85.

———. 2009b. "Waiting for Global Warming: Climate Politics in Australia—Perspectives from Above and Below." In *Waiting*, edited by Ghassan Hage, 135–49. Melbourne, Australia: Melbourne University Press.

Baer, Hans A., and Merrill Singer. 2009. *Global Warming and the Political Ecology of Health: Emerging Crises and Systemic Solutions*. Walnut Creek, CA: Left Coast Press.

Baer, Hans A., Merrill Singer, and Ida Susser. 2003. *Medical Anthropology and the World System: A Critical Perspective*. 2nd ed. Westport, CT: Praeger.

Baer, Paul, Sivan Kartha, Tom Athanasiou, and Eric Kemp-Benedict. 2009. "The Greenhouse Development Rights Framework: Drawing Attention to Inequality within Nations in the Global Climate Policy Debate." *Development and Change* 40:1121–38.

Baran, Paul A., and Paul M. Sweezy. 1966. *Monopoly Capital: An Essay on the American Economic Order*. New York: Monthly Review Press.

Barbier, Edward B. 2010. *A Global Green New Deal: Rethinking the Economic Recovery*. Cambridge: Cambridge University Press.

Barnett, Jon, and John Campbell. 2010. *Climate Change and Small Island States: Power, Knowledge and the South Pacific*. London: Earthscan.

Barnosky, Anthony D. 2009. *Heatstroke: Nature in an Age of Global Warming*. Washington, DC: Island Press.

Bartley, Tim, and Albert Bergesen. 1997. "World-System Studies of the Environment." *Journal of World System Research* 3:369–80.

Bauman, Zygmunt. 2011. *Collateral Damage: Social Inequalities in a Global Age*. London: Polity.

Beder, Sharon. 2004. *Global Spin: The Corporate Assault on Environmentalism*. Melbourne, Australia: Scribe Publications.

Behringer, Wolfgang. 2010. *A Cultural History of Climate*. Cambridge, UK: Polity.

Bello, Walden. 2007. "Climate Change Flap at the G8," ZNet. July 7. http://www.zcommunications.org/climate-change-flap-at-the-g8-by-walden-bello.

———. 2009. *The Food Wars*. London: Verso.

Berger, John J. 2000. *Beating the Heat: Why and How We Must Combat Global Warming*. Berkeley, CA: Berkeley Hills Books.

Betts, R. A., M. Collins, D. I. Hemming, C. D. Jones, J. A. Lowe, and M. G. Sanderson. 2011. "When Could Global Warming Reach 4°C?" *Philosophical Transactions of the Royal Society A* 369:67–84.

Betzien, Jody. 2011. "Carbon Price Could Delay Real Climate Action." *Green Left Weekly*. March 23, 10.

Biel, Robert. 2006. "The Interplay between Social and Environmental Degradation in the Development of the International Political Economy." *Journal of World-Systems Research* 12, no. 1:109–47.

Black, Brian C., and Gary J. Weisel. 2010. *Global Warming*. Santa Barbara, CA: Greenwood.

Black, Edwin. 2006. *Internal Combustion: How Corporations and Governments Addicted the World to Oil and Derailed the Alternatives*. New York: St. Martin's Griffin.

Black, William R. 2007. "Sustainable Solutions for Freight Transfer." In *Globalized Freight Transport*, edited by T. R. Leinbach and C. Capineri, 189–216. Aldershot, UK: Edward Elgar.

———. 2010. Sustainable Transportation: Problems and Solutions. New York: Guilford Press.

Bocock, Robert. 1993. *Consumption*. London: Routledge.

Bodley, John H. 2012. *Anthropology and Contemporary Human Problems.* 6th ed. Lanham, MD: AltaMira Press.

———. 2008. *Anthropology and Contemporary Human Problems*, 5th ed. Walnut Creek, CA: AltaMIra Press.

Bond, Patrick. 2009. "Carbon Charade." In *People First Economics*, edited by David Ransom and Vanessa Bird, 211–16. Oxford: World Changing.

———. 2010. "Maintaining Momentum after Copenhagen's Collapse: Seal the Deal or 'Seattle' the Deal?" *Capitalism Nature Socialism* 21, no. 1:14–27.

Boswell, Terry. 2004. "American World Empire or Declined Hegemony." *Journal of World Systems Research* 10:516–24.

Boswell, Terry, and Christopher Chase-Dunn. 2000. *The Spiral of Capitalism and Socialism*. Boulder, CO: Lynne Rienner.

Brennan, T. 2003. *Globalisation and Its Terrors*. London: Routledge.

Brinton, Maurice. 1975. *The Bolsheviks and Workers' Control.* London: Black & Red.

Brockington, Dan, Rosaleen Duffy, and Jim Igoe. 2008. *Nature Unbound: Conservation, Capitalism and the Future of Protected Areas.* London: Earthscan.

Brohé, Arnaud, Nick Eyre, and Nicholas Howarth. 2009. *Carbon Markets: An International Business Guide*. London: Earthscan.

Brown, Bob, and Peter Singer. 1996. *The Greens*. Melbourne, Australia: Text Publishing Company.

Brown, Hillary. 2010. "Toward Zero-Carbon Buildings." In *The Post-Carbon Reader: Managing the 21st Century's Sustainability Crises*, edited by Richard Heinberg and Daniel Lerch, 322–34. Healdsburg, CA: Watershed Media.

Brown, Lester R. 2009. *Plan B 4.0: Mobilizing to Save Civilization*. New York: W. W. Norton & Company.

Brunnengräber, Achim. 2006. "Political Economy of the Kyoto Protocol." In *Coming to Terms with Nature: Socialist Register 2007*, edited by Leon Panitch and Colin Leys, 213–30. London: Merlin Press.

Brunner, Steffan. 2008. "Understanding the Policy Change: Multiple Streams and Emissions Trading in Germany." *Global Environmental Change* 18:501–7.

Bryner, Gary C. 2001. *Gaia's Wager: Environmental Movements and the Challenge of Sustainability*. Lanham, MD: Rowman & Littlefield.

Bullard, Nicola. 2010. "Climate Debt: A Subversive Political Strategy." Climate Justice Now. April 18. http://www.climate-justice-now.org/climate-debt-a-subversive-political-strategy.

Bunting, Andrea Maree. 2003. "The Social Shaping of Wind Power in Australia." PhD diss. School of Social Sciences, Media, and Communication, University of Wollongong, New South Wales, Australia.

Burgmann, Verity. 2003. *Power, Profit and Protest: Australian Social Movements and Globalisation*. 2nd ed. Crows Nest, New South Wales: Allen & Unwin.

Burman, Stephen. 2007. *The State of the American Empire: How the USA Shapes the World*. London: Earthscan.

Busby, Joshua W. 2007. *Climate Change and National Security: An Agenda for Action*. Council Special Report No. 32. New York: Council on Foreign Relations.

Butler, Simon. 2010a. "BP Restricted Media, Can't Restrict Oil." *Green Left Weekly*, June 23, 17.

———. 2010b. "Gillard: Same Old Spin on Climate." *Green Left Weekly*. July 7, 2010.

Caffentzis, George. 2010. "'Everything Must Change So That Everything Can Remain the Same': Reflections on Obama's Energy Plan." In *Sparking a Worldwide Energy Revolution: Social Struggles in the Transition to a Post-Petrol World*, edited by Kolya Abramsky, 564–68. Oakland, CA: AK Press.

Cahill, Michael. 2010. *Transport, Environment and Society*. Maidenhead, Berkshire, UK: Open University Press.

Caldicott, Helen. 2006. *Nuclear Power Is Not the Answer to Global Warming or Anything Else.* Carlton, Victoria: Melbourne University Press.

———. 2009. *If You Love This Planet: A Plan to Save the Earth.* Rev. ed. New York: W. W. Norton & Company.

Camilleri, Joseph A., and Jim Falk. 2010. *Worlds in Transition: Evolving Governance across a Stressed Planet*. Cheltenham, UK: Edward Elgar.

Carbon Trade Watch. 2005. *Hoodwinked in the Hothouse: The G8, Climate Change and Free-Market Environmentalism.* Amsterdam: Transnational Institute.

Carrier, J. G., and D. V. I. Macleod. 2005. "Bursting the Bubble: The Socio-Cultural Context of Eco-Tourism." *Journal of the Royal Anthropological Institute* 11:315–34.

Carter, Neil. 2007. *Politics of the Environment: Ideas, Actions, Policy*, 2nd ed. Cambridge, UK: Cambridge University Press.

Castells, Manuel. 2004. *End of Millennium.* 2nd ed. Malden, MA: Blackwell.

———. 2009. *Communication Power*. Oxford: Oxford University Press.

Chafe, Zoe. 2008. "Air Travel Reaches New Heights." In *Vital Signs 2007–2008: The Trends That Are Shaping Our Future*, edited by Linda Starke, 70–71. New York: W. W. Norton & Company.

Chandler, Jo. 2011. *Feeling the Heat*. Melbourne, Australia: Melbourne University Press.

Charman, Karen. 2008. "False Starts and False Solutions: Current Approaches in Dealing with Climate Change." *Capitalism Nature Socialism* 19:29–47.

Chase-Dunn, Christopher. 2005. "Social Evolution and the Future of World Society." *Journal of World-Systems Research* 11, no. 2:171–92.

Chivers, Danny. 2009. "Climate Choices." In *People First Economics*, edited by David Ransom and Vanessa Baird, 193–209. Oxford, UK: World Changing.

Chossudovsky, Michel. 2004. "The Ultimate Weapon of Mass Destruction: 'Owning the Weather' for Military Use." *Global Research* 27 (September), www.globalresearch.ca.

Christoff, Peter. 2010. "Cold Climate in Copenhagen: China and the United States at COP15." *Environmental Politics* 19:637–56.

Clark, Brett, and Richard York. 2005. "Carbon Metabolism: Global Capitalism, Climate Change, and Biospheric Rift." *Theory and Society* 34:391–428.

CNA Corporation. n.d. "National Security and the Threat of Climate Change." www.cna.org

Coelho, Ricardo. 2011. "The EU Carbon Market Scam." *Green Left Weekly*. March 9.

Cooke, Sophie. 2010. "Leave It in the Ground—the Growing Global Struggle against Coal." In *Sparking a Worldwide Energy Revolution: Social Struggles in the Transition to a Post-Petrol World*, edited by Kolya Abramsky, 424–38. Oakland, CA: AK Press.

Cooper, Richard N. 2007. "Alternatives to Kyoto: The Case for a Carbon Tax." In *Architecture for Agreement: Addressing Global Climate Change in the Post-Kyoto World*, edited by Joseph E. Aldy and Robert N. Stavins, 105–15. Cambridge: Cambridge University Press.

Council of the European Union. 2008. *Climate Change and International Security*. Report from the Commission and the Secretary-General/High Representative to the European Council, 7249/08, Brussels, March 3.

Cowie, Jonathan. 1998. *Climate Change and Human Change: Disaster or Opportunity?* London: Parthenon Publishing Group.

Crate, Susan A., and Mark Nuttall, eds. 2009. *Anthropology and Climate Change: From Encounters to Actions*. Walnut Creek, CA: Left Coast Press.

Craven, Greg. 2009. *What's the Worst That Could Happen? A Rational Response to the Climate Change Debate*. New York: Perigee Book.

D'Amato, Paul. 2006. *The Meaning of Marxism*. Chicago: Haymarket Books.

Dale, Gareth. 2007. "'On the Menu or at the Table': Corporations and Climate Change." *International Socialism* 116:117–38.

Dallas, Nick. 2009. *Green Business Basics: 24 Lessons for Meeting the Challenges of Global Warming*. New York: McGraw-Hill.

Dauvergne, Peter. 2005. "Dying of Consumption: Accidents or Sacrifices of Global Mortality?" *Global Environmental Politics* 5(3): 35–47

Dauvergne, Peter, and Jane Lister. 2011. *Timber*. London: Polity.

Davidson, Christopher M. 2008. *Dubai: The Vulnerability of Success*. London: Hurst & Company.

Davis, David Howard. 2007. *Ignoring the Apocalypse: Why Planning to Prevent Environmental Catastrophe Goes Astray*. Westport, CT: Praeger.

Dawson, Brian, and Matt Spannagle. 2009. *The Complete Guide to Climate Change*. London: Routledge.

De Cauter, Lieven. 2008. "The Mad Max Phase." In *Urban Climate Change Crossroads*, edited by Richard Plunz and Maria Paola Sutto, 111–17. Surrey, UK: Ashgate.

Death, Carl. 2011. "Summit Theatre: Exemplary Governmentality and Environmental Diplomacy in Johannesburg and Copenhagen." *Environmental Politics* 20:1–19.

Deloitte. 2009. "Energy Security—America's Best Defense." Deloitte. November 10. www.deloitte.com.

Dessler, Andrew, and Edward A. Parson. 2010. *The Science and Politics of Global Climate Change: A Guide to the Debate.* 2nd ed. Cambridge: Cambridge University Press.
Diamond, Jared. 2005. *Collapse: How Societies Choose to Fail or Succeed.* New York: Penguin.
Dicken, Peter. 2003. *Global Shift: Reshaping the Global Economic Map in the 21st Century.* 4th ed. New York: Guilford Press.
Dilworth, Craig. 2010. *Too Smart for Our Own Good: The Ecological Predicament of Humankind.* Cambridge: Cambridge University Press.
Dimento, Joseph F. C., and Pamela Doughman. 2007. "Climate Change: How the World Is Responding." In *Climate Change: What It Means for Us, Our Children, and Our Grandchildren,* edited by Joseph F. C. DiMento and Pamela Doughman, 101–38. Cambridge, MA: MIT Press.
Dimitrov, Radoslav S. 2010. "Inside Copenhagen: The State of Climate Governance." *Global Environmental Politics* 10, no. 2 (May):18–24.
Drake, Frances. 2000. *Global Warming: The Science of Climate Change.* London: Arnold.
Driesen, David M. 2010. "Neoliberal Instrument Choice." In *Economic Thought and U.S. Climate Change Policy,* edited by David M. Driesen, 129–49. Cambridge, MA: MIT Press.
Eckersley, Robyn. 2006. "From the Liberal to the Green Democratic State: UpholdingAutonomy and Sustainability." *International Journal of Innovation and Sustainable Development* 1: 266–83.
Editors. 2002. "U.S. Imperial Ambitions and Iraq." *Monthly Review* 54, no. 7 (December), 1–13, http://monthlyreview.org/2002/12/01/u-s-imperial-ambitions-and-iraq.
——. 2011a. "Backing Tax on Workers No Way to Stop Abbott." *Solidarity* 33 (March–April 2011), http://www.solidarity.net.au/33/backing-tax-on-workers-no-way-to-stop-abbott.
——. 2011b. "Deniers, Pretenders and Real Climate Action." *Green Left Weekly*. March 20. http://www.greenleft.org.au/node/47044.
Ekan, Ivar. 2007. In Sweden, its global warming vs. big heavy cars. *The New YorkTimes,* July 5
Ellerman, A. Denny, Frank J. Convery, and Christian de Perthuis. 2010. *Pricing Carbon: The European Union Emissions Trading Scheme.* Cambridge: Cambridge University Press.
Emanuel, Kamala. 2009. "What Alternatives to the CPRS." *Green Left Weekly*. March 25.
Emmerson, Charles. 2010. *The Future History of the Arctic.* London: Bodley Head.
Energy Watch Group. 2006. Uranium Resources and Nuclear Energy. EWG-Series No 1/2007. Energy Watch Group. http://www.energy-

watchgroup.org/fileadmin/global/pdf/EWG_Report_Uranium_3-12-2006ms.pdf.

Engdahl, F. William. 2007. "China and the USA in a New Cold War over Africa's Oil Riches: Darfur? It's the Oil, Stupid." *Global Research*. May 20. http://www.globalresearch.ca/index.php?context=va&aid=5714.

Ensor, Jonathan. 2011. *Uncertain Futures: Adapting Development to a Changing Climate*. Warwickshire, UK: Practical Action Publishing.

Environment Victoria. 2009. "The People's Climate White Paper: A Community Action Plan for Our Climate." Environment Victoria. September. http://environmentvictoria.org.au/library/community-action-plan-our-climate.

Environmental Defense Fund. 2007. Cars by the Numbers: Statistics on Automobiles and Their Global Warming Contribution. www.edf.org, accessed May 9, 2008.

Epstein, Paul, and Christine Rogers. 2004. *Inside the Greenhouse: The Impacts of CO_2 and Climate Change on Public Health in the Inner City*. Boston: Center for Health and the Global Environment.

Fagan, Brian. 1999. *Floods, Famines, and Emperors: El Nino and the Fate of Civilization*. Cambridge: Cambridge University Press.

———. 2008. *The Great Warming: Climate Change and the Rise and Fall of Civilizations*. New York: Bloomsbury Press.

Falk, Jim, and A. Brownlow. 1989. *Greenhouse Challenge: What's to Be Done?* Penguin: Melbourne.

Farley, John W. 2008. "The Scientific Case for Modern Anthropogenic Global Warming." *Monthly Review* (July–August): 68–90.

Farndon, John. 2008. *India Booms: The Breathtaking Development and Influence of Modern India*. London: Virgin Books.

Faulks, Keith. 1999. *Political Sociology: A Critical Introduction*. Edinburgh: Edinburgh University Press.

Finley, Mary Lou. 2007. "Shaping the Movement." In *Ignition: What You Can Do to Fight Global Warming and Spark a Movement*, edited by Jonathan Isham and Sissel Waage, 33–56. Washington, DC: Island Press.

Firat, A. Fuat, and Nikhilesh Dholakia. 1998. *Consuming People: From Political Economy to Theaters of Consumption*. London: Routledge.

Fitz, Don. 2007. "What's Possible in the Military Sector? Greater Than 100% Reduction in Greenhouse Gases." ZNet. April 30. http://www.zcommunications.org/whats-possible-in-the-military-sector-by-don-fitz.

Flannery, Tim. 2005. *The Weather Makers*. New York: Atlantic Monthly Press.

——. 2010. *Here on Earth: An Argument for Hope*. Melbourne, Australia: Text Publishing.

Flavin, Christopher, and Robert Engelman. 2009. "The Perfect Storm." In *State of the World 2009: Confronting Climate Change*, edited by Linda Starke, 5–29. London: Earthscan.

Forest, James J. F., and Matthew V. Sousa. 2006. *Oil and Terrorism in the New Gulf: Framing U.S. Energy and Security Policies for the Gulf of Guinea*. Lanham, MD: Lexington Books.

Foster, John Bellamy. 2000. *Marx's Ecology*. New York: Monthly Review Press.

——. 2005. "Organizing Ecological Revolution." *Monthly Review* (October): 1–10.

——. 2009a. "A Failed System: The World Crisis of Capitalist Globalization and Its Impact on China." *Monthly Review* (March): 1–23.

——. 2009b. *The Ecological Revolution: Making Peace with the Planet*. New York: Monthly Review Press.

——. 2010. "Why Ecological Revolution?" *Monthly Review*, January, pp. 1–64.

——. 2011. "Capitalism and Degrowth—an Impossibility Theorem." *Monthly Review* (January): 26–33.

Foster, John Bellamy, Brett Clark, and Richard York. 2009. "The Midas Effect: A Critique of Climate Change Economics." *Development and Change* 40:1085–97.

——. 2010. "Capitalism and the Curse of Energy Efficiency: The Return of the Jevons Paradox." *Monthly Review* (November): 1–12.

Frank, Christine. 2009. "The Bankruptcy of Capitalist Solutions to the Climate Crisis." *Capitalism Nature Socialism* 20:32–43.

Friel, Howard. 2010. *The Lomborg Deception: Setting the Record Straight about Global Warming*. New Haven, CT: Yale University Press.

Friends of the Earth Australia. 2008. "Open Submission to Garnaut on the Emissions Trading Scheme." Garnaut Climate Change Review. April 17. http://www.garnautreview.org.au/ca25734e0016a131/WebObj/D0850341ETSSubmission-FriendsoftheEarth/$File/D08%2050341%20%20ETS%20Submission%20-%20Friends%20of%20the%20Earth.pdf.

——. 2011. "Hard-Wired Path to Carbon Trading Goes the Wrong Direction." Coalition for a Safe Climate. February 24. http://www.safeclimate.org.au/node/405.

Friends of the Earth Europe. 2010a. *The Cancun Agreement Not Worth Cheering*. Brussels: FoE Europe.

——. 2010b. *The EU Emissions Trading System: Failing to Deliver*. Brussels: FoE Europe.

Froelich, Paul. 1972. *Rosa Luxemburg: Ideas in Action*. London: Pluto Press.
Frumkin, H. 2001. "Beyond Toxicity: Human Health and the Natural Environment." *American Journal of Preventive Medicine* 20, no 3:234–40.
Fugletvedt, Jan, et al. 2008. "Climate Forcing from the Transport Sectors." *Proceedings of the North American Academy of Science* 105(2): 454–58.
Gallopin, Gilberto, Al Hammond, Paul Raskin, and Rob Swat. 1997. *Global Scenarios and Human Choice: A Resource Paper of the Global Scenario Group*. Stockholm, Sweden: Stockholm Environment Institute.
Gardner, Gary. 2008a. "Aluminium Production Continues Upwards." In *Vital Signs 2007–2008: The Trends That Are Shaping Our Future*, edited by Linda Starke, 58–59. New York: W. W. Norton & Company.
———. 2008b. "Roundwood Production Up." In *Vital Signs 2007–2008: The Trends That Are Shaping Our Future*, edited by Linda Starke, 62–63. New York: W. W. Norton & Company.
Garnaut, Ross. 2008. *The Garnaut Climate Change Review*. Cambridge: Cambridge University Press.
Gautier, Catherine. 2008. *Oil, Water, and Climate: An Introduction*. Cambridge: Cambridge University Press.
Gelbspan, Ross. 2004. *Burning Point: How Politicians, Big Oil and Coal, Journalists, and Activists Are Fueling the Climate Crisis—and What We Can Do to Avert Disaster*. New York: Basic Books.
German Advisory Council on Global Change. 2008. *World in Transition: Climate Change as a Security Risk*. London: Earthscan.
Gershon, David. 2009. *Social Change 2.0: A Blueprint for Reinventing Our World*. West Harley, NY: High Point/Chelsea Green.
Giddens, Anthony. 2009. *The Politics of Climate Change*. Cambridge, UK: Polity.
Gilbert, Richard, and Anthony Perl. 2010a. *Transport Revolutions: Moving People and Freight without Oil*. Rev. ed. London: Earthscan.
———. 2010b. "Transportation in the Post-Carbon World." In *The Post Carbon Reader: Managing the 21st Century's Sustainability Crises*, edited by Richard Heinberg and Daniel Lerch, 347–260. Healdsburg, CA: Watershed Media.
Gilbertson, Tamara, and Oscar Reyes. 2009. *Carbon Trading: How It Works and Why It Fails*. Critical Currents No. 7. Uppsala, Sweden: Dag Hammarskjöld Foundation.
Gilding, Paul. 2011. *The Great Disruption: How the Climate Crisis Will Transform the Global Economy*. London: Bloomsbury.
Glantz, Michael H. 2003. *Climate Affairs: A Primer*. Washington, DC: Island Press.
Global Carbon Project. 2010. Carbon Budget Highlights. www.globalcarbonproject.org/carbonbudget/10/hl-full.htm.

———. 2011. Carbon Budget 2010. Report released December 5, 2011. www.globalcarbonproject.org.

Global Footprint Network. 2010. "Ecological Footprint and Biocapacity, 2007."Global Footprint Network. October 13, 2010. http://www.footprintnetwork.org/ismages/uploads/2010_NFA_data_tables.pdf.

Godrej, Dinyar. 2006. *The No-Nonsense Guide to Climate Change*. Oxford: New Internationalist Publications.

Gonzalez, George A. 2005. "Urban Sprawl, Global Warming and the Limits of Ecological Modernisation." *Environmental Politics* 14:344–62.

———. 2008. "Urban Sprawl, Climate Change, Oil Depletion, and Eco-Marxism." In *Political Theory and Global Climate Change*, edited by Steve Vanderheiden, 153–76. Cambridge, MA: MIT Press.

Goodall, Chris. 2007. *How to Live a Low-Carbon Life: The Individual's Guide to Stopping Climate Change*. London: Earthscan.

Goodstein, Eban. 2007. "Your Mission: Focus the Nation." In *Ignition: What You Can Do to Fight Global Warning and Spark a Movement*, edited by Jonathan Isham and Sissel Waage, 157–68. Washington, DC: Island Press.

Gore, Al. 2006. *An Inconvenient Truth: The Planetary Emergency of Global Warming and What We Can Do About It*. London: Rodale.

———. 2007. *Earth in the Balance: Forging a New Common Purpose*. 2nd ed. London: Earthscan (1st ed. in 1992).

———. 2009. *Our Choice: A Plan to Solve the Climate Crisis*. London: Bloomsbury.

Gorz, Andre. 1973. *Socialism and Revolution*. Garden City, NY: Anchor.

Graeber, David. 2009. *Direct Action: An Ethnography*. Oakland, CA: AK Press.

Green, Jim. n.d. "Nuclear Power: No Solution to Climate Change." Paper prepared for Friends of the Earth Australia, the Australian Conservation Foundation, Greenpeace Australia Pacific, the Medical Association for the Prevention of War, the Public Health Association of Australia, and Climate Action Network Australia.

Grimes, Peter, and Jeffrey Kentor. 2003. "Exporting the Greenhouse: Foreign Capital Penetration and CO_2 Emissions 1980–1996." *Journal of World-Systems Research* 9, no. 2 (Summer): 261–75.

Gulick, John. 2011. "The Long Twentieth Century and Barriers to China's Hegemonic Accession." *Journal of World-Systems Research* 17, no 1:4–38.

Hahnel, Robin. 2008. "Participatory Economics and the Environment: Chris Spannos Interviews Robin Hahnel." In *Real Utopia: Participatory Society for the 21st Century*, edited by Chris Spannos, 56–74. Oakland, CA: AK Press.

Hamilton, Clive. 2007. *Scorcher: The Dirty Politics of Climate Change*. Melbourne, Australia: Black Incorporated.

———. 2009. "The Time for Radical Activism." Speech delivered at the Climate Action Summit, Australian National University, January 31.

Hamilton, Clive, and Richard Denniss. 2005. *Affluenza: When Too Much Is Never Enough*. Crows Nest, New South Wales: Allen &Unwin.

Hammond, Allen. 1998. *Which World? Scenarios for the 21st Century*. Washington, DC: Island Press.

Hansen, James. 2007. "Climate Catastrophe." *New Scientist* 298:30–34.

———. 2009. *Storms of My Grandchildren: The Truths about the Coming Climate Catastrophe and Our Last Chance to Save Humanity*. New York: Bloomsbury Press.

Hansen, Lara J., and Jennifer R. Hoffman. 2011. *Climate Savvy: Adapting Conservation and Resource Management to a Changing World*. Washington, DC: Island Press.

Hardt, Michael, and Antonio Negri. 2009. *Commonwealth*. Cambridge, MA: Harvard University Press.

Hardy, John T. 2003. *Climate Change: Causes, Effects, and Solutions*: Hoboken, NJ: John Wiley & Sons.

Hare, W. L. 2009. "A Safe Landing for the Climate." *In* Linda Starke, ed., *State of the World: Confronting Climate*, 13–29. London: Earthscan.

Harris, Paul G. 2010. *World Ethics and Climate Change: From International to Global Justice*. Edinburgh: Edinburgh University Press.

Harvey, David. 2000. *Spaces of Hope*. Edinburgh: Edinburgh University Press.

———. 2011. *The Enigma of Capital and the Crises of Capitalism*. London: Profile Books.

Hassol, S. 2004. *Impacts of a Warming Arctic: Arctic Climate Impact Assessment*. Cambridge: Cambridge University Press.

Heinberg, Richard. 2006. *The Oil Depletion Protocol: A Plan to Avert Oil, Wars, Terrorism and Economic Collapse*. Gabriola Island, BC: New Society Publishers

———.2010. *Peak Everything: Waking Up to the Century of Declines*. Gabriola Island, BC: New Society Publishers.

Henson, Robert. 2006. *The Rough Guide to Climate Change: The Symptoms, the Science, the Solutions*. London: Penguin.

Hertsgaard, Mark. 2011. *Hot: Living through the Next Fifty Years on Earth.* Boston: Houghton Mifflin Harcourt.

Hetherington, Renee, and Robert G. B. Reid. 2010. *The Climate Connection: Climate Change and Modern Human Evolution.* Cambridge: Cambridge University Press.

High Representative and the European Commission to the European Council. 2008. *Climate Change and International Security.* March 14.

Hillman, Mayer. 2004. *How We Can Save the Planet.* London: Penguin.

Hofrichter, Richard. 2000. "Introduction: Critical Perspectives on Human Health and the Environment." In *Reclaiming the Environmental Debates: The Politics of Health in a Toxic Culture,* edited by Richard Hofrichter, 1–15. Cambridge, MA: MIT Press.

Holloway, John. 2002. *Change the World without Taking Power.* London: Pluto.

Honore, Carl. 2004. *Praise of Slow: How a Worldwide Movement Is Challenging the Cult of Speed.* London: Orion.

Hopkins, Rob. 2008. *The Transition Handbook: Creating Local Sustainable Communities beyond Oil Dependency.* Sydney: Finch Publishing.

———. 2010. "What Can Communities Do?" In *The Post Carbon Reader: Managing the 21st Century's Sustainability Crisis,* edited by Richard Heinberg and Daniel Lerch, 442–51. Healdsburg, CA: Watershed Media.

Hornborg, Alf. 2001. *The Power of the Machine: Global Inequalities of Economy, Technology, and Environment.* Walnut Creek, CA: AltaMira Press.

Hossay, Patrick. 2006. *Unsustainable: A Primer for Global Environmental and Social Justice.* London: Zed Books.

Howard, Roger. 2009. *The Arctic Gold Rush: The New Race for Tomorrow's Natural Resources.* New York: Continuum.

Hughes, D. H. 2000. *Ecology in Ancient Civilizations.* Albuquerque: University of New Mexico Press.

Hughes, J. David. 2010. "Hydrocarbons in North America." In *The Post-Carbon Reader: Managing the 21st Century's Sustainability Crises,* edited by Richard Heinberg and Daniel Lerch, 211–28. Headleburg, CA: Watershed Media.

Hulme, Mike. 2009. *Why We Disagree about Climate Change: Understanding Controversy, Inaction and Opportunity.* Cambridge: Cambridge University Press.

Intergovernmental Panel on Climate Change (IPCC). 2007a. *A Report of Working Group I of the Intergovernmental Panel on Climate Change: Summary for Policymakers.* IPCC. http://www.ipcc.ch/pdf/assessment-report/ar4/wg1/ar4-wg1-spm.pdf.

———. 2007b. *Contribution of Working Group II to the Fourth Assessment Report of the Intergovernmental Panel on Climate Change*. Cambridge: Cambridge University Press.

International Energy Agency (IEA). 2010. *CO_2 Emissions from Fuel Combustion Highlights*. Paris: IEA.

Isham, Jonathan, and Sissel Waage. 2007. "Igniting Action for a New Movement." In *Ignition: What You Can Do to Fight Global Warming and Spark a Movement*, edited by Jonathan Isham and Sissel Waage, 11–20. Washington, DC: Island Press.

Isla, Ana. 2009. "Who Pays for the Kyoto Protocol? Selling Oxygen and Selling Sex in Costa Rica." In *Eco-Sufficiency and Global Justice: Women Write Political Ecology*, edited by Ariel Salleh, 199–217. North Melbourne, Victoria: Spinifex Press.

Jackson, Tim. 2009. "Prosperity without Growth? The Transition to a Sustainable Economy." Sustainable Development Commission. March 30. http://www.sd-commission.org.uk/data/files/publications/prosperity_without_growth_report.pdf.

Jacobs, Michael. 2010. "Europe's Radical Carbon Choice." *Inside Story*. October 6. http://inside.org.au/europe-carbon-choice.

Jami, Dahr. 2010. "Tar Sands—A Toxic Nightmare." *Green Left Weekly*. June 23, 16.

Jarman, Melanie. 2007. *Climate Change*. London: Pluto Press.

Jennaway, Megan. 2006. "Reflections on an Economy of Fluids: Ecology, History and Health in East Timor." Paper presented at the Australian Anthropology Society Conference, James Cook University, Cairns, Australia, September 26–29.

Johansen, Bruce E. 2006. *Global Warming in the 21st Century*. Vol. 2: *Melting Ice and Warming Seas*. Westport, CT: Praeger.

Johnston, Barbara Rose. 2011. "Anthropological Voice on the Fukushima Disaster." *News: A Publication of the Society for Applied Anthropology* 22, no. 2:6–10.

Jones, Bart. 2007. *Hugo! The Hugo Chavez Story from Mud Hut to Perpetual Revolution*. Hanover, NH: Steerforth Press.

Jones, Mark. 2005. "Battle of Titans." In *The Final Energy Crisis*, edited by Andrew McKillop and Sheila Newman, 105–15. London: Pluto.

Jull, Peter. 2009–2010. "Indigenous Sovereignty in the Northwest Passage." *Arena Magazine* 103:45.

Juris, Jeffrey S. 2009. *Networking Futures: The Movements against Corporate Globalization*. Durham, NC: Duke University Press.

Kaplan, Robert D. 2010. *Monsoon: The Indian Ocean and the Battle for Supremacy in the 21st Century*. Collingwood, Victoria, Australia: Black, Incorporated.

Karbuz, Sohbet. 2007. "US Military Pains." *Energy Bulletin* (archived February 17).

Karlenzig, Warren. 2010. "The Death of Sprawl: Designing Urban Resilience for the Twenty-First Century Resource and Climate Crises." In *The Post Carbon Reader: Managing the 21st Century's Sustainability Crises*, edited by Richard Heinberg and Daniel Lerch, 295–313. Healdsburg, CA: Watershed Media.

Karliner, Joshua. 2000. "The Globalization of Corporate Culture and Its Role in the Environmental Crisis." In *Reclaiming the Environmental Debate: The Politics of Health in a Toxic Culture*, edited by Richard Hofrichter, 178–97. Cambridge, MA: MIT Press.

Kasarda, John D., and Greg Lindsay. 2011. *Aerotropolis: The Way We'll Live Next*. New York: Farrar, Straus and Giroux.

Kaye, John. n.d. "Rudd's White Flag of Surrender on Climate Change." John Kaye. http://johnkaye.org.au/campaigns/energy/emissions.

Kennedy, Donald. "Preserving the Conditions of Life." 2006. In *Science Magazine's State of the Planet 2006–2007*, edited by Donald Kennedy and Editors of Science, 39–48. Washington, DC: Island Press.

Kiely, Ray. 2007. *The New Political Economy of Development: Globalization, Imperialism, and Hegemony*. New York: Palgrave MacMillan.

Kinmouth, William. 2004. *Climate Change: A Natural Hazard*. Brentwood, Essex, UK: Multi-Science Publishing.

Kister, Chad. 2005. *Arctic Melting: How Global Warming Is Destroying One of the World's Largest Wilderness Areas*. Monroe, ME: Common Courage Press.

Klare, Michael T. 2001. *Resource Wars: The New Landscape of Global Conflict*. New York: Henry Holt and Company.

———. 2003. "Blood for Oil: The Bush-Cheney Energy Strategy." *Socialist Register* 40: http://socialistregister.com/index.php/srv/article/view/5816.

———. 2004. "Blood for Oil: The Bush-Cheney Energy Strategy." In *Socialist Register 2004—The New Imperial Challenge*, edited by Leo Panith and Colin Leys, 166–83. London: Merlin Press.

———. 2007. Iraq and Climate Change. *Foreign Policy in Focus* (December 7).

———. 2008. *Rising Powers, Shrinking Planet: The New Geopolitics of Energy*. New York: Metropolitan Books.

Kohler, Brian. 2010. "Sustainability and Just Transition in the Energy Industries." In *Sparking a Worldwide Energy Revolution: Social Struggles in the Transition to a Post-Petrol World*, edited by Kolya Abramsky, 569–76. Oakland, CA: AK Press.

Kolko, Gabriel. 2006. *The Age of War: The United States Confronts the World.* Delhi: Gyan Sagar Publications.

Korten, David. 2001. *When Corporations Rule the World.* 2nd ed. SF: Kumarian Press.

Kovel, Joel. 2007. *The Enemy of Nature: The End of Capitalism or the End of the World?* 2nd ed. London: Zed Books.

———. 2008. "Ecosocialism, Global Justice, and Climate Change." *Capitalism Nature Socialism* 19:4–14.

Krane, Jim. 2009. *City of Gold: Dubai and the Dream of Capitalism.* New York: St. Martin's Press.

Kruse, Jane. 2010. "The End of One Danish Windmill Cooperative." In *Sparking a Worldwide Energy Revolution: Social Struggles in the Transition to a Post-Petrol World,* edited by Kolya Abramsky, 523–24. Oakland, CA: AK Press.

Ladd, Brian. 2008. *Autophobia: Love and Hate in the Automotive Age.* Chicago: University of Chicago.

Lafargue, Paul. 1999. *The Right to Be Lazy.* Ardmore, PA: Fifth Season Press.

Langholz, Jeffrey, and Kelly Turner. 2003. *You Can Prevent Global Warming (and Save Money!).* Kansas City, MO: Andrews McMeel Publishing.

Lebowitz, Michael. 2010. *The Socialist Alternative: Real Human Development.* New York: Monthly Review Press.

Leggett, Jeremy. 2005. *Half Gone: Oil, Gas, Hot Air and the Global Energy Crisis.* London: Portobello Books.

Leonard, Mark. 2008. *What Does China Think?* London: HarperCollins.

Levins, Richard. 2005. "How Cuba Is Going Ecological." *Capitalism Nature Socialism* 16(3):7–25.

Li, Minqi. 2008. "Climate Change, Limits to Growth and the Imperative for Socialism." *Monthly Review* (July–August): 51–67.

———. 2009. "Capitalism, Climate Change and the Transition to Sustainability: Alternative Scenarios for the US, China and the World." *Development and Change* 40:1039–61.

Linden, Eugene. 2006. *Winds of Change: Climate, Weather, and the Destruction of Civilizations.* New York: Simon & Schuster.

Little, Amanda. 2009. *Power Trip: From Oil Wells to Solar Cells—Our Ride to the Renewable Future.* London: Harper Press.

Liu, Yingling. 2008. "Steel Production Soars." In *Vital Signs 2007–2008: The Trends That Are Shaping Our Future,* edited by Linda Starke, 56–57. New York: W. W. Norton & Company.

Lohmann, Larry. 2009. "Climate as Investment." *Development and Change* 40:1063–83.

Lo, Alex Y. 2010. Active Conflict or Passive Coherence? The Political Economy of Climate Change in China. *Environmental Politics* 19:1012–17.

Lowy, Michael. 2006. "Eco-Socialism and Democratic Planning." In *Coming to Terms with Nature*, edited by Leo Panitch and Colin Leys, 294–309. London: Merlin Press.

Lohan, Tara. 2009. "Towns Rush to Make Low-Carbon Transition." *Yes! Magazine*. December 1. http://www.yesmagazine.org/issues/climate-action/towns-rush-to-make-low-carbon-transition.

Lohmann, Larry. 2006. *Carbon Trading: A Critical Conversation on Climate Change, Privatisation and Power*. Development Dialogue No. 28. Uppsala, Sweden: Dag Hammarskjoeld Centre.

Lomborg, Bjorn. 2007. *The Skeptical Environmentalist: Measuring the Real State of the World*. Cambridge: Cambridge University Press.

———. 2010. *Cool It: The Skeptical Environmentalist's Guide to Global Warming*. Cambridge: Cambridge University Press.

Longhai, Shen. 2008. "Correct Choices for China: Energy Conservation, a Cyclic Economy, and a Conservation-Minded Society." In *Global Warming: Looking beyond Kyoto*, edited by Ernesto Zedillo, 215–23. Washington, DC: Brookings Institution Press.

Lorimer, Doug. 1997. *The Collapse of "Communism" in the USSR: Its Causes and Significance*. Chippendale, New South Wales: Resistance Books.

Lovell, Bryan. 2010. *Challenged by Carbon: The Oil Industry and Climate Change*. Cambridge: Cambridge University Press.

Lovelock, James. 2006. *The Revenge of Gaia*. London: Penguin Books.

———. 2009. *The Vanishing Face of Gaia: A Final Warning*. Camberwell, Victoria: Allen Lane.

Low, Nicholas, Brendan Gleeson, Ray Green, and Darko Radovic. 2005. *The Greencity: Sustainable Homes, Sustainable Suburbs*. Sydney: University of New South Wales Press.

Lowe, Ian. 2005. *Living in the Hothouse: How Global Warming Affects Australia*. Melbourne, Australia: Scribe.

———. 2010. *A Voice of Reason: Reflections on Australia*. Brisbane: University of Queensland Press.

Luke, Tim W. 2008. "Climatologies as Social Critique: The Social Construction/Creation of Global Warming, Global Dimming, and Global Cooling." In *Political Theory and Global Climate Change*, edited by Steve Vanderheiden, 121–52. Cambridge, MA: MIT Press.

Luxemburg, Rosa. [1913] 1951. *The Accumulation of Capital*. London: Routledge and K. Paul.

Lynas, Mark. 2004. *High Tide: News from a Warming World*. London: Flamingo.
———. 2007. *Six Degrees: Our Future on a Hotter Planet*. London: Fourth Estate.
Lyster, Rosemary. 2010. "Reducing Emissions from Deforestation and Degradation: The Road to Copenhagen." In *In the Wilds of Climate Law*, edited by Rosemary Lyster, 97–142. Bowen Hills, Queensland: Australian Academic Press.
MacKay, David J. C. 2009. *Sustainable Energy—without Hot Air*. Cambridge: Cambridge University Press.
Maddison, Angus. 2001. *The World Economy: A Millennial Perspective*. Paris: Organization for Economic Cooperation and Development (OECD).
Magdoff, Fred. 2008. "The Political Economy and Ecology of Biofuels." *Monthly Review* 60, no. 3 (July–August): 34–50.
———. 2011. "Ecological Civilization." *Monthly Review* 62, no. 8 (January): 1–25.
Magdoff, Fred, and John Bellamy Foster. 2010. "What Every Environmentalist Needs to Know about Capitalism." *Monthly Review* 61, no. 10 (March): 1–30.
Mamdani, Mahomood. 1972. *The Myth of Population Control: Family, Caste, and Class in an Indian Village*. New York: Monthly Review Press.
Mann, Scott. 2010. *Bioethics in Perspective: Corporate Power, Public Health and Political Economy*. Port Melbourne, Victoria: Cambridge University Press.
Marcuse, Herbert. 1991. *One Dimensional Man*. London: Routledge.
Markandya, Anil. 2009. "Can Climate Change Be Reversed under Capitalism?" *Development and Change* 40:1139–52.
Maslin, Mark. 2004. *Global Warming: A Very Short Introduction*. Oxford: Oxford University Press.
———. 2009. "Hot or Cold Future?" In *The Complete Ice Age: How Climate Change Shaped the World*, edited by Brian Fagan, 206–31. London: Thames & Hudson.
Mate, Janos, Kert Davies, and David Kanter. 2009. "The Risks of Other Greenhouse Gases." In *State of the World: Confronting Climate Change*, edited by Linda Starke, 52–55. London: Earthscan.
McAuley, Ian. 2010. "Taxing the Miners' Uncommonly Large Profit." *Dissent* (Spring): 20–25.
McChesney, Robert W., and John Bellamy Foster. 2010. "Capitalism, the Absurd System: A View from the United States." *Monthly Review*

62, no. 2 (June). http://monthlyreview.org/2010/06/01/capitalism-the-absurd-system-a-view-from-the-united-states.
McCluney, Ross. 2005. "Renewable Energy Limits." In *The Final Energy Crisis*, edited by Andrew McKillop and Sheila Newman, 153–75. London: Pluto Press.
McCright, Aaron M., and Riley E. Dunlap. 2000. "Challenging Global Warming as a Social Problem: An Analysis of the Conservative Movement's Counter-Claims." *Social Problems* 47:499–522.
McKeown, Alice, and Gary Gardner. 2009. *Climate Change Reference Guide*. Washington, DC: Worldwatch Institute.
McKibben, Bill. 2007. *Fight Global Warming Now: The Handbook for Taking Action in Your Community*. New York: Holt Paperbacks.
———. 2010. *Eaarth: Making a Life on a Tough New Planet*. Melbourne, Australia: Black Inc.
McKillop, Andrew. 2005. "The Myth of Decoupling." In *The Final Energy Crisis*, edited by Andrew McKillop with Sheila Newman, 197–216. London: Pluto.
McMichael, Anthony J. 1993. *Planetary Overload: Global Environmental Change and the Health of the Species*. Cambridge: Cambridge University Press.
———. 2001. *Human Frontiers, Environments and Disease*. Cambridge: Cambridge University Press.
McQuaig, Linda. 2004. *Crude, Dude: War, Big Oil and the Fight for the Planet*. Toronto: Doubleday Canada.
Meadows, Donnella H., Dennis L. Meadows, Jorgen Randers, and William W. Behrens III. 1972. *The Limits to Growth*. New York: Universe, 126.
Meadows, Donnella, Jorgen Randers, and Dennis Meadows. 2004. *Limits to Growth: The 30-Year Update*. London: Earthscan.
Meagher, Gabrielle, and Shaun Wilson. 2008. "Richer, but More Unequal: Perceptions of Inequality in Australia, 1987-2005." *Journal of Australian Political Economy*, 61: 221–40.
Mees, Paul. 2010. *Transport for Suburbia: Beyond the Automobile Age*. London: Earthscan.
Melucci, Alberto. 1980. "The New Social Movements." *Social Science Information* 19:196–226.
Metz, Bert. 2010. *Controlling Climate Change*. Cambridge: Cambridge University Press.
Middle, Peter. 2010. "How Novels Can Contribute to Our Understanding of Climate Change." In *History at the End of the World: History, Climate Change and the Possibility of Closure*, edited by Mark Levene, Rob Johnson, and Penny Roberts, 218–33. UK: Humanities-Ebooks.

Middleton, Paul. 2007. *A Brief Guide to the End of Oil*. London: Robinson.
Milanovic, Branko. 2002. "True World Income Distribution, 1988 and 1993: First Calculation Based on Household Surveys Alone," *Economic Journal* 112, no. 476 (January): 51–92.
Milner, Andrew. 1994. "On the Beach: Apocalyptic Hedonism and the Origins of the Postmodern." In *Australian Popular Culture*, edited by Ian Carven, 190–204. Cambridge: Cambridge University Press.
Monbiot, George. 2006. *Heat: How to Stop the Planet Burning*. Camberwell, Victoria, Australia: Allen Lane.
Montgomery, Scott L. 2010. *The Powers That Be: Global Energy for the Twenty-First Century and Beyond*. Chicago: University of Chicago Press.
Moore, Jess. 2011. "Carbon Price: The Gas Agenda." *Green Left Weekly*. June 29, 2011.
Moran, Emilio F. 2010. *Environmental Social Science: Human-Environment Interactions and Sustainability*. Malden, MA: Wiley-Blackwell.
Morgan, Gareth, and John McCrystal. 2009. *Poles Apart: Beyond the Shouting and Who's Right about Climate Change?* Melbourne, Australia: Scribe.
Moser, Susanne C. 2009. "Communicating Climate Change and Motivating Civic Action: Renewing, Activating, and Building Democracies." In *Changing Climates in North American Politics: Institutions, Policymaking and Multilevel Governance*, edited by Henrik Selin and Stacey D. Van Deever, 283–302. Cambridge, MA: MIT Press.
Mueller, Tadzio, and Alexis Passadakis. 2010. "Another Capitalism Possible? From World Economic Crisis to Green Capitalism." In *Sparking a Worldwide Energy Revolution: Social Struggles in the Transition to a Post-Petrol World*, edited by Kolya Abramsky, 554–63. Oakland, CA: AK Press.
Mumford, Lewis. 1934. *Technics and Civilization*. New York: Harcourt.
Murphy, K. 2008. "Carbon Trading by 2012." *The Age*, June 1.
Murphy, Pat. 2008. *Plan C: Community Survival Strategies for Peak Oil and Climate Change*. Gabriola Island, BC: New Society Publishers.
Myers, Norman, and Jennifer Kent. 2004. *The New Consumers: The Influence of Affluence on the Environment*. Washington, DC: Island Press.
NASA. 2010. "The Russian Heat Wave of 2010." www.esrl.nooa.gov/noaa.gov/psd/csi/events/2010/russian heat wave/prelim.html.
NASA Goddard Institute for Space Studies (GISS). 2010. "How Warm Was Summer 2010?," NASA GISS. September 30. http://www.giss.nasa.gov/research/news/20100930.
National Development and Reform Commission. 2007. "China's National Climate Change Program." China Climate Change Info-Net.

http://www.ccchina.gov.cn/WebSite/CCChina/UpFile/File188.pdf.
———. 2009. China's Policies and Actions for Addressing Climate Change—The Progress Report 2009. Beijing.
National Geographic. 2004. "Global Warming: Bulletins from a Warmer World." *National Geographic* 206, no. 3 (September).
Neale, Jonathan. 2008. *Stop Global Warming: Change the World*. UK: Bookmarks Publications.
New, Mark, Diana Liverman, Heike Schroder, and Kevin Anderson. 2011. "Four Degrees and Beyond: The Potential for a Global Temperature Increase of Four Degrees and Its Implications." *Philosophical Transactions of the Royal Society* 369:6–19.
Newell, Peter, and Matthew Paterson. 2010. *Climate Capitalism: Global Warming and the Transformation of the Global Economy*. Cambridge: Cambridge University Press.
Newman, Peter. 2009. "Transport Opportunities: Towards a Resilient City. In John O'Brien, ed., *Opportunities Beyond Carbon: Looking for a Sustainable World*. 98–115. Melbourne: Melbourne University Press.
———. 2010. "Responding to Oil Vulnerability and Climate Change in Our Cities." In *Managing Climate Change: Papers from the Greenhouse 2009 Conference*, edited by Imogen Jubb, Paul Holper, and Wenjce Cai, 177–83. Collingwood, Victoria: CSIRO Publishing.
Newman, Peter, and Isabella Jennings. 2008. *Cities as Sustainable Ecosystems: Principles and Practices*. Washington, DC: Island Press.
Newman, Peter, and Jeffrey Kenworthy. 1999. *Sustainability and Cities: Overcoming Automobile Dependence*. Washington, DC: Island Press.
Nitzan, Jonathan, and Shimson Bichler. 2002. *The Global Political Economy of Israel*. London: Pluto Press.
Nordhaus, William. 2008. *A Question of Balance: Weighing the Options on Global Warming Policies*. New Haven, CT: Yale University Press.
Norgaard, Kari Marie. 2011. *Living in Denial: Climate Change, Emotions, and Every Day Life*. Cambridge, MA: MIT Press.
O'Brien, Karen L., and Robin M. Leichenko. 2009. "Double Exposure: Assessing the Impacts of Climate Change within the Context of Economic Globalization." In *The Earthscan Reader on Adaptation to Climate Change*, edited by E. Lisa F. Schipper and Ian Burton, 323–42. London: Routledge.
O'Riordan, Timothy, and Andrew Jordan. 1999. "Institutions, Climate Change and Cultural Theory: Towards a Common Analytical Framework." *Global Environmental Change* 9:81–93.
Oceransky, Sergio. 2010. "Fighting the Enclosure of Wind: Indigenous Resistance to the Privatization of the Wind Resource in Southern

Mexico." In *Sparking a Worldwide Energy Revolution: Social Struggles in the Transition to a Post-Petrol World,* edited by Kolya Abramsky, 504–22. Oakland, CA: AK Press.

Officer, Charles, and Jake Page. 2009. *When the Planet Rages: Natural Disasters, Global Warming, and the Future of the Earth.* Oxford: Oxford University Press.

Onaran, Özlem. 2010. "The Crisis of Capitalism in Europe, West and East." *Monthly Review* 62, no. 5 (October): , 18–33,http://monthlyreview.org/2010/10/01/the-crisis-of-capitalism-in-europe-west-and-east.

Panayotakis, Costas. 2006. "Working More, Selling More, Consuming More: Capitalism's 'Third Contradiction.'." In *Coming to Terms with Nature,* edited by Leo Panitch and Colin Leys, 254–72. London: Merlin Press.

Panitch, Leo. 2008. *Renewing Socialism: Democracy, Strategy, and Imagination.* Boulder, CO: Westview Press.

Parkinson, Stuart. 2007. "Guns and Global Warming: War, Peace and the Environment." Web version of presentation given at the Network for Peace AGM, London, February, Scientists for Global Responsibility, http://www.sgr.org.uk/resources/guns-and-global-warming-war-peace-and-environment.

Paskal, Cleo. 2010. *Global Warring: How Environmental, Economic, and Political Crises Will Redraw the World Map.* New York: Palgrave Macmillan.

Pattberg, Philipp, and Johannes Stripple. 2010. "Agency in Global Climate Governance: Setting the Stage." In *Global Climate Governance beyond 2012: Architecture, Agency and Adaptation,* edited by Frank Biermann, Philip Pattberg, and Fariborz Zelli, 137–45. Cambridge: Cambridge University Press.

Patterson, Matthew. 2007. *Automobile Politics: Ecology and Cultural Political Economy.* Cambridge, UK: Cambridge University Press.

Peak Oil News. 2006. "Military Oil Usage Statistics." Peak Oil News. October 15. http://www.peak-oil-news.info/military-oil-usage-statistics (accessed March 8, 2008).

Pearce, Fred. 2006. *The Last Generation: How Nature Will Take Her Revenge for Climate Change.* London: Transworld Publishers.

———. 2007. *With Speed and Violence: Why Scientists Fear Tipping Points in Climate Change.* Boston: Beacon Press.

———. 2010. *The Climate Files: The Battle for the Truth about Global Warming.* London: Guardian Books.

Pearse, Guy. 2007. *High and Dry: Climate Change and the Selling of Australia's Future.* Melbourne, Australia: Viking.

Pearse, Rebecca. 2010/2011. "Making a Market? Contestation and Climate Change." *Journal of Australian Political Economy* 66:166–98.

Perlmutter, Daniel D., and Robert L. Rothstein. 2011. *The Challenge of Climate Change: Which Way Now?* Oxford, UK: Wiley-Blackwell.

Petermann, Anne, and Orin Langelle. 2010. "What Really Happened in Copenhagen." *Z Magazine*. February. http://www.zcommunications.org/what-really-happened-in-copenhagen-by-anne-petermann.

Pittock, A. Barrie. 2008. "Ten Reasons Why Climate Change May Be More Severe Than Projected." In *Sudden and Disruptive Climate Change: Exploring the Real Risks and How We Can Avoid Them*, edited by Michael C. MacCracken, Frances Moore, and John C. Topping Jr., 11–27. London: Earthscan.

———. 2009. *Climate Change: The Science, Impacts and Solutions*, 2nd ed. Collingwood, Victoria, Australia: CSIRO Publishing.

Plimer, Ian. 2009. *Heaven and Earth: Global Warming: The Missing Science*. Ballan, Victoria: Connor Court Publishing.

Ponting, Clive. 2007. *A New Green History of the World: The Environment and the Collapse of Great Civilizations*. London: Vintage.

Pooley, Eric. 2010. *The Climate War: True Believers, Power Brokers, and the Fight to Save the Earth*. New York: Hyperion.

Porter, Richard C. 1999. *Economics at the Wheel: The Costs and Drivers*. San Diego: Academic Press.

Poston, Dudley L., Jr., and Leon F. Bouvier. 2010. *Population and Society: An Introduction to Demography*. Cambridge: Cambridge University Press.

Potter, Robert B., Tony Binns, Jennifer A. Elliott, and David Smith. 2008. *Geographies of Development: An Introduction to Development Studies*. Essex, UK: Pearson Prentice Hall.

Prinn, Ronald G., and John M. Reilly. 2011. "Climate Change Today: A Look at New Lessons from Ongoing Global Change Research." MIT Joint Program on the Science and Policy of Global Change, April.

Raskin, Paul, Tariq Banuri, Gilberto Gallopin, Pablo Gutman, Al Hammond, Robert Kates, and Rob Swart. 2002. *Great Transition: The Promise and the Lure of the Times Ahead*. Stockholm: Global Scenario Report.

Reay, Dave. 2005. *Climate Change Begins at Home: Life on the Two-Way Street of Global Warming*. London: Macmillan.

Register, Richard. 2001. *Ecocities: Building Cities in Balance with Nature*. Berkeley, CA: Berkeley Hills Books.

Renner, Michael. 1997. "Environmental and Health Effects of Weapons Production, Testing, and Maintenance." In Barry S. Levy and Victor W. Sidel, eds., *War and Public Health,* 117–36. New York: Oxford University Press.

———. 2008. "Vehicle Production Rises Sharply." In *Vital Signs 2007–2008: The Trends That Are Shaping Our Future,* edited by Worldwatch Institute, 66–67. New York: W. W. Norton & Company.

Resistance. 1999. *Environment, Capitalism and Socialism.* Chippendale, New South Wales: Resistance Books.

Richter, Burton. 2010. *Beyond Smoke and Mirrors: Climate Change and Energy in the 21st Century.* Cambridge: Cambridge University Press.

Rickards, Lauren, and Karlie Tucker. 2009. "Challenges for Australian Agriculture." In *Climate Change for Young and Old,* edited by Helen Sykes, 84–101. Sydney: Future Leaders.

Roberts, J. Timmons, Peter E. Grimes, and Jodie L. Manale. 2003. "Social Roots of Global Environmental Change: A World-Systems Analysis of Carbon Dioxide Emissions." *Journal of World-Systems Research* 9, no. 2:277–315.

Roberts, J. Timmons, and Bradley C. Parks. 2007. *A Climate of Injustice: Global Inequality, North-South Politics, and Climate Policy.* Cambridge, MA: MIT Press.

Rogelj, Joeri, Julia Nabel, Claudine Chen, William Hare, Kathleen Markmann, Malte Meinshausen, Michiel Schaeffer, Kirsten Macy, and Niklas Hohne. 2010. "Copenhagen Accord Pledges Are Paltry." *Nature* 464 (April 22): 1126–28.

Rogers, Heather. 2010. *Green Gone Wrong: How Our Economy Is Undermining the Environmental Revolution.* New York: Scribner.

Roncoli, C., and J. Magistro. 2000. "Global Science, Local Practice: Anthropological Dimensions of Climate Variability." *Practicing Anthropology* 22, no 4:2–5.

Rosa, Eugene A., and Thomas Dietz. 2010. "Global Transformations: PaSSAGE to a New Ecological Era." In *Human Footprints on the Global Environment: Threats to Sustainability,* edited by Eugene A. Rosa, Andreas Dielmann, Thomas Dietz, and Carlo Jaeger, 1–45. Cambridge, MA: MIT Press.

Ruddiman, William F. 2005. *Plows, Plagues and Petroleum: How Humans Took Control of Climate.* Princeton, NJ: Princeton University Press.

Rumble, Eric. 2006. "Explaining National Climate Change Policies." *Global Environmental Change* 5:235–49.

Ruppert, Michael C. 2009. *Confronting Collapse: The Crisis of Energy and Money in a Post Peak Oil World—a 25-Point Program for Action.* White River Junction, VT: Chelsea Green Publishing.

Russell, Joshua Kahn. 2009. "Climate Justice and Coal's Funeral Procession." *Z-Magazine*. May. http://www.zcommunications.org/climate-justice-and-coals-funeral-procession-by-joshua-kahn-russell.

Rutledge, Ian. 2006. *Addict to Oil: America's Relentless Drive for Energy Security*. London: I.B. Tauris.

Ruyle, Eugene. 1977. "A Socialist Alternative for the Future." In *Cultures of the Future*, edited by Magorah Maruyma and Arthur M. Harkins, 613–28. The Hague: Mouton.

Sachs, Jeffrey D. 2008. *Common Wealth: Economics for a Crowded Planet*. New York: Allen Lane.

Safe Climate Australia. n.d. "Al Gore launches Safe Climate Australia." Safe Climate Australia. http://www.safeclimateaustralia.org/media-and-links.

———. 2009. Media Release. Safe Climate Transition Plan to Be Developed for Australia. July 7, www.safeclimateaustra

———. 2012. The Australian Safe Climate Transition Plan. www.safeclimateaustralia,org, accessed March 24.

Sale, Peter F. 2011. *Our Dying Planet: An Ecologist's View of the Crisis We Face*. Berkeley: University of California Press.

Salleh, A. 2010. "Embodied Materialism in Action." In Canavan, Gerry, Lisa Klarr, and Ryan Vu, eds. Special Issue on Ecology and Ideology. Polygraph, (22): 183–99, www.duke.edu/web/polygraph/cfp.html. lia.org.

Salleh, Ariel, ed. 2009. *Eco-Sufficiency and Global Justice: Women Write Political Ecology*. Melbourne, Australia: Spinifex Press.

———. 2010. "Embodied Materialism in Action." In Gerry Canavan, Lisa Klarr, and Ryan Vu, eds. Special Issue on Ecology and Ideology. *Polygraph*, (22): 183–99, www.duke.edu/web/polygraph/cfp.html.

Sanders, Barry. 2009. *The Green Zone: The Environmental Costs of Militarism*. Oakland, CA: AK Press.

Saunders, Ewan. 2010. "Climate Action Summit 2010: A Step Forward." *Green Left Weekly*. March 24, 11.

Sawin, Janet L. 2008. "Carbon Emissions Continue Unrelenting Rise." In *Vital Signs 2007–2008: The Trends That Are Shaping Our Future*, edited by Worldwatch Institute, 42–43. New York: W. W. Norton & Company.

Sawin, Janet L., and Ishani Mukherjee. 2008. "Fossil Fuel Use Up Again." In *Vital Signs 2007–2008: The Trends That Are Shaping Our Future*, edited by Worldwatch Institute, 32–33. New York: W. W. Norton & Company.

Schaefer, Andreas, John B. Heywood, Henry D. Jacoby, and Ian A. Waitz. 2009. *Transportation in a Climate-Constrained World.* Cambridge, MA: Cambridge University Press.

Scherr, Sara J., and Sajal Sthapit. 2009. "Farming and Land Use to Cool the Planet." In *State of the World 2009,* edited by Linda Starke, 30–49. London: Earthscan.

Schor, Juliet. 1992. *The Overworked American: The Unexpected Decline of Leisure.* Boulder, CO: Basic Books.

———. 2010. *Plenitude: The New Economics of True Wealth.* New York: Penguin.

Schreuder, Yda. 2009. *The Corporate Greenhouse: Climate Change Policy in a Globalizing World.* London: Zed Books.

Schwartz, Peter, and Doug Randall. 2003. "An Abrupt Climate Change Scenario and Its Implications for United States National Security." Global Business Network. October. http://www.gbn.com/articles/pdfs/Abrupt%20Climate%20Change%20February%202004.pdf.

Schwartzman, David. 2009. "Ecosocialism or Ecocatastrophe?" *Capitalism Nature Socialism* 20, no. 1: 6–33..

Scott-Cato, Molly, and Jean Hillier. 2010. "How Could We Study Climate-Related Social Innovation? Applying Deleuzean Philosophy to Transition Towns." *Environmental Politics* 19:869–87.

Shannon, Thomas R. 1996. *An Introduction to the World-System Perspective.* 2nd ed. Boulder, CO: Westview Press.

Shapley, Dan. 2007. "Global Warming and War: 6 Flashpoints." The Daily Green, December 6. http://www.thedailygreen.com/environmental-news/latest/national-security-47120607.

Shaw, Christopher. 2010. "Dangerous Limits: Climate Change and Modernity." In *History at the End of the World: History, Climate Change and the Possibility of Closure,* edited by Mark Levene, Rob Johnson, and Penny Roberts, 95–111. UK: Humanities-Ebooks.

Shearer, Christine. 2011. *Kivalina: A Climate Change Story.* Chicago: Haymarket Books.

Shearman, David, and Joseph Wayne Smith. 2007. *The Climate Change Challenge and the Failure of Democracy.* Westport, CT: Praeger.

Sheehan, Bill, and Helen Spiegelman. 2010. "Climate Change, Peak Oil, and the End of Waste." In *The Post Carbon Reader: Managing the 21st Century's Sustainability Crises,* edited by Richard Heinberg and Daniel Lerch, 363–81. Healdsburg, CA: Watershed Media.

Sherman, Howard J. 1995. *Reinventing Marxism.* Baltimore: Johns Hopkins University.

Shiva, Vandana. 2008. *Soil Not Oil: Environmental Justice in an Age of Climate Crisis*. Boston: South End Press.

Sikor, T., J. Stahl, T. Enters, J. C. Ribot, N. M. Singh, W. D. Sunderlin, and E. Wollenberg, 2010. "REDD-plus, Forest People's Rights and Nested Climate Governance." *Global Environmental Change* 20, no. 3 (August):423–25.

Sim, Stuart. 2010a. *The Carbon Footprint Wars: What Might Happen If We Retreat from Globalization?* Edinburgh: Edinburgh University Press.

———. 2010b. *The End of Modernity: What the Financial and Environmental Crisis Is Really Telling Us*. Edinburgh: Edinburgh University Press.

Simms, Andrew. 2009. *Ecological Debt: Global Warming and the Wealth of Nations*. 2nd ed. London: Pluto Press.

Singer, S. Fred, and Dennis T. Avery. 2007. *Unstoppable Global Warming Every 1,500 Years*. Lanham, MD: Rowman & Littlefield.

Skjærseth, Jon Birger, and Tora Skodvin. 2009. *Climate Change and the Oil Industry: Common Problem, Varying Strategies*. Manchester: Manchester University Press.

Sklair, Leslie. 2006. *Globalization: Capitalism & Its Alternatives*. Oxford, UK: Oxford University Press.

Slater, Don. 1997. *Consumer Culture and Modernity*. Cambridge, UK: Polity Press.

Smith, Brice. 2006. *Insurmountable Risks: The Dangers of Using Nuclear Power to Combat Global Climate Change*. Takoma Park, MD: IEER Press.

Smith, Dan, and Janani Vivekananda. 2007. "A Climate of Conflict: The Links between Climate Change, Peace and War." *International Alert*. November.

Smith, Gar. 1990–1991. "How Fuel-Efficient Is the Pentagon? Military's Oil Addiction." *Earth Island Journal* (Winter). http://www.envirosagainstwar.org/know/read.php?itemid+593m (accessed June 12, 2008).

Smith, Kevin. 2007. *The Carbon Neutral Myth: Offset Indulgences for Your Climate Sins*. Amsterdam: Carbon Trade Watch.

Smith, Laurence C. 2011. *The New North: The World in 2050*. London: Profile Books.

Smith, Neil. 2010. *Uneven Development: Nature, Capital and Production of Space*. 3rd ed. London: Verso.

Solty, Ingar. 2008. "The Historic Significance of the New German Left Party," translated by Eric Canepa. *Socialism and Democracy* 22:1–34.

Spaargaren, Gert, and Arthur P. J. Mol. 2008. "Greening Global Consumption: Redefining Politics and Authority." *Global Environmental Change* 18:350–59.

Spence, Christopher. 2005. *Global Warming: Personal Solutions for a Healthy Planet*. New York: Palgrave.

Sperling, Daniel. 2010. *Two Billion Cars: Driving Toward Sustainability*. New York: Oxford University Press.

Speth, James Gustave. 2008. *The Bridge at the Edge of the World: Capitalism, the Environment, and Crossing from Crisis to Sustainability*. New Haven, CT: Yale University Press.

Spratt, David, and Phillip Sutton. 2008a. *Climate Code Red: The Case for a Sustainability Emergency*. Fitzroy: Friends of the Earth Australia.

———. 2008b. *Climate Code Red: The Case for Emergency Action*. Melbourne, Australia: Scribe.

Starr, Amory. 2003. *Naming the Enemy: Anti-Corporate Movements Confront Globalization*. New York: Zed Books.

Steinman, David. 2007. *Safe Trip to Eden: Ten Steps to Save Planet Earth from Global Warming Meltdown*. New York: Thunder's Mouth Press.

Stern, Nicholas. 2007. *The Economics of Climate Change: The Stern Review*. Cambridge: Cambridge University Press.

———. 2009. *A Blueprint for a Safer Planet: How to Manage Climate Change and Create a New Era of Progress and Prosperity*. London: Bodley Head.

Stern, Pamela R. 2010. *Daily Life of the Inuit*. Santa Barbara, CA: Greenwood Press.

Stilwell, Frank. 1992. *Understanding Cities and Regions*. Leichhardt, New South Wales: Pluto Australia.

———. 2000. *Changing Track: A New Political Economic Direction for Australia*. Sydney: Pluto Press.

Storm, Servaas. 2009. "Capitalism and Climatic Change: Can the Invisible Hand Adjust the Natural Thermostat?" *Development and Change* 40:1011–38.

Surin, Kenneth. 2009. *Freedom Not Yet: Liberation and the Next World Order*. Durham, NC: Duke University Press.

Sutter, Christoph, and Juan Carlos Parreno. 2007. "Does the Current Clean Development Mechanism (CDM) Deliver Its Sustainable Development Claim? An Analysis of Officially Registered CDM Projects." *Climatic Change* 84:75–90.

Symon, Carolyn, Lelani Arris, and Bill Heal, eds. 2004. *Arctic Climate Impact Assessment*. Cambridge: Cambridge University Press.

Synott, John. 2004. *Global and International Studies: Transdisciplinary Perspectives*. Southbank, Victoria: Social Science Press Australia.

Tanuro, Daniel. 2009. "Climate Crisis: 21st Century Socialists Must Be Ecosocialists." In *The Global Fight for Climate Justice: Anticapitalist Responses to Global Warming and Environmental Destruction*, edited by Ian Angus, 237–82. London: Resistance Books.

Taylor, Graeme. 2008. *Evolution's Edge: The Coming Collapse of Our World*. Gabriola Island, BC: New Society Publishers.

Thomas, Amy. 2011. "Carbon Price Plan: Shovel Billions to Business," *Solidarity* 32 (February).

Tickell, Oliver. 2008. *Kyoto2: How to Manage the Global Greenhouse*. London: Zed Books.

Tiffen, Rodney, and Ross Gittens. 2009. *How Australia Compares*, 2nd ed. Cambridge, UK: Cambridge University Press.

Tippet, Gary, and Mark Russell. 2011. "Hotter, Colder, Wetter: It's a New World of Extremes." *The Age*, January 16, 9.

Tokar, Brian. 2009a. "Mobilization for Climate Justice Open Letter to the Grassroots: Help Organize for Urgent Action on Climate Change." ZNet. May 18. http://zcommunications.org/mobilization-for-climate-justice-open-letter-to-the-grassroots-help-organize-for-urgent-action-on-climate-change-by-brian-tokar.

———. 2009b. "Toward Climate Justice: Can We Turn Back from the Abyss." *Z Magazine* 22, no. 9 (September): http://www.zcommunications.org/toward-climate-justice-by-brian-tokar.

Tomain, Joseph P. 2010. "Dirty Energy Policy." In *Economic Thought and U.S. Climate Change Policy*, edited by David M. Driesen, 45–60. Cambridge, MA: MIT Press.

Trainer, Ted. 1989. *Developed to Death: Rethinking Third World Development*. London: Green Print.

———. 1991. "A Green Perspective on Inequality." In *Inequality in Australia: Slicing the Cake—the Social Justice Collective*, edited by Jan O'Leary and Rachel Sharp, 105–24. Port Melbourne, Victoria: William Heinemann Australia.

———. 1995. *The Conserver Society: Alternatives for Sustainability*. London: Zed Books.

———. 1998. *Saving the Environment: What It Will Take*. Sydney: University of New South Wales Press.

———. 2009. The Transitions Towns Movement: Its Huge Significance and a Friendly Criticism. (We) Can Do Better. July 30. http://candobetter.net/node/1439.

———. 2011. *The Transition to a Sustainable and Just World*. Canterbury, New South Wales: Envirobook.

Turner, George. 1987. *The Sea and the Summer*. London: Faber and Faber.

Turner, Martyn, and Brian O'Connell. 2001. *The Whole World's Watching: Decarbonizing the Economy and Saving the World*. Chichester, UK: John Wiley & Sons.

Turshen, Meredith. 2004. "The Politics of Oil in Africa." *The Scholar and Feminist Online* 2.2 (winter): http://barnard.edu/sfonline/reverb/printmtu.htm.

United Nations. 2004. *World Urbanization Prospects: The 2003 Revision*. United Nations. http://www.un.org/esa/population/publications/wup2003/WUP2003Report.pdf.

———. 2005. *World Population Prospects: The 2004 Revision*. United Nations. February 24. http://www.un.org/esa/population/publications/WPP2004/2004Highlights_finalrevised.pdf.

United Nations Development Programme. 1999. *Human Development Report 1999*. New York: Oxford University Press.

———. 2007. *Human Development Report 2007/2008—Fighting Climate Change/Human Solidarity in a Divided World*. New York: Palgrave Macmillan.

United Nations Environmental Programme. 2009. "Green New Deal: An Update for the G20 Pittsburgh Summit." September, www.unep.org/pdf/G20_policy_Final.pdf.

———. 2010. *Assessing the Environmental Impact of Consumption and Production*. UNEP. http://www.unep.org/resourcepanel/Portals/24102/PDFs/PriorityProductsAndMaterials_Report.pdf.

United Nations Food and Agriculture Organization (FOA). 2007. *State of the World's Forests*. Rome: FOA.

US Department of Energy, Energy Information Administration (EIA). 2008. *Annual Energy Review*. http://www.eia.gov/FTPROOT/multifuel/038408.pdf.

US Department of Transportation (DOT). 2010. *Transportation's Role in Reducing U.S. Greenhouse Gas Emissions: Report to Congress*. US DOT. April. http://ntl.bts.gov/lib/32000/32700/32779/DOT_Climate_Change_Report_-_April_2010_-_Volume_1_and_2.pdf.

US Energy Information Administration (EIA). "US Imports by Country of Origin." EIA. June 29, 2009. http://tonto.eia.doe.gov.

Vanderheiden, Steven. 2008. "Introduction." In *Political Theory and Global Climate*, edited by Steven Vanderheiden, xi–xxiv. Cambridge, MA: MIT Press.

Vaze, Prashant. 2009. *The Economical Environmentalist: My Attempt to Live a Low-Carbon Life and What It Costs*. London: Earthscan.

Victor, David G. 2004. *Climate Change: Debating America's Policy Options.* New York: Council on Foreign Relations.

Vogt, Kristina A., Toral Patel-Weynard, Maura Shelton, Daniel J. Vogt, John C. Gordon, Cal Mukumoto, Asep S. Suntani, and Patricia A. Road. 2010. *Sustainability Unpacked: Food, Energy and Water for Resilient Environments and Societies.* London: Earthscan.

Vrommen, Ariadne, Katherine Gelber, and Anika Gauja. 2009a. *Powerscape: Direction for Australia.* Sydney: Pluto Press.

———. 2009b. *Powerscape: Contemporary Australian Politics.* Crows Nest, New South Wales: Allen & Unwin.

Wackernagel, M., and W. Rees. 1996. *Our Ecological Footprint: Reducing Human Impact on the Earth.* Gabriola Island, BC: New Society Publishers.

Wade, Robert. 2003. *Real World Economic Outlook.*

Wagar, W. Warren. 1992. *A Short History of the Future.* Chicago: University of Chicago Press.

Wall, Derek. 2010a. *The Rise of the Green Left: Inside the Worldwide Ecosocialist Movement.* London: Pluto Press.

———. 2010b. *The No-Nonsense Guide to Green Politics.* Oxford: New Internationalist.

Wallerstein, Immanuel. 1979. *The Capitalist World-Economy: Essays.* New York: Cambridge University Press.

———. 1998. *Utopistics: Or Historical Choices of the Twenty-First Century.* New York: New Press.

———. 2007. "Climate Disasters: Three Obstacles to Doing Anything." Commentary No. 205. Fernand Braudel Center. March 15. http://www2.binghamton.edu/fbc/archive/205en.htm.

———. 2011a. "Structural Crisis in the World-System: Where Do We Go from Here?" *Monthly Review* 62, no. 10 (March): 31–39.

———. 2011b. "The World Consequences of U.S. Decline." Commentary No. 311. Fernand Braudel Center. August 15. http://www2.binghamton.edu/fbc/commentaries/archive-2011/311en.htm.

Wallis, Victor. 2006. "Vision and Strategy: Questioning the Subsistence Perspective." *Capitalism Nature Socialism* 17, no. 4:38–43.

———. 2010. "Beyond 'Green Capitalism.'" *Monthly Review* 61, no. 9 (February): 32–48.

Ward, Peter D. 2007. *Under a Green Sky: Global Warming, the Mass Extinctions of the Past, and What They Can Tell Us about Our Future.* New York: HarperCollins.

———. 2010. *The Flooded Earth: Our Future in a World without Ice Caps.* New York: Basic Books.

Warren, Kay B. 2006. "Perils and Promises of Engaged Anthropology: Historical Transitions and Ethnographic Dilemmas." In *Engaged Observer: Anthropology, Advocacy and Activism*, edited by Victoria Sanford and Asale Angel-Ajani, 213–28. Brunswick, NJ: Rutgers University Press.

Watts, Jonathan. 2010. *When a Billion Chinese Jump: How China Will Save Mankind—or Destroy It*. London: Faber and Faber.

Weart, Spencer R. 2003. *The Discovery of Global Warming*. Cambridge, MA: Harvard University Press.

Webb, Dave. 2007. "Thinking the worst—The Pentagon Report" In *Clearing the Pathwys to Survival—The State, Ourselves, and Climate Change*, edited by David Cromwell and Mark Levene 59–81. London: Pluto.

Wen, Dale. 2010. "Climate Change, Energy and China: Technology, Market and Beyond." In *Sparking a Worldwide Energy Revolution: Social Struggles in the Transition to a Post-Petrol World*, edited by Kolya Abramsky, 130–54. Oakland, CA: AK Press.

Wettenhall, Roger. 1965. "Public Ownership in Australia." *Political Quarterly* 36:426–40.

Whelan, James, with Pru Gell and Holly Hammond. 2008. *Building the Climate Movement Online: Action Research Project*. The Change Agency. September. http://www.thechangeagency.org/_dbase_upl/BldgClimMvntOnline.pdf.

Whipple, Tom. 2010. "Peak Oil and the Great Recession." In *The Post Carbon Reader: Managing the 21st Century's Sustainability Crisis*, edited by Richard Heinberg and Daniel Lerch, 247–55. Healdsburg, CA: Watershed Media.

White, David, and Dick Nicholls. 2009. "Socialist Alliance and the Movement against Global Warming." *Alliance Voices: Socialist Alliance Discussion Bulletin and National Newsletter*, March, http://alliancevoices.blogspot.com/2009/03/around-branchessocialist-alliance-in.html.

White, Leslie A. 2008. *Modern Capitalist Culture*, edited by Burton J. Brown, Benjamin Urish, and Robert L. Carneiro. Walnut Creek, CA: Left Coast Press.

Wilkinson, R., and K. Pickett. 2009. *The Spirit Level*. London: Allen Lane.

World Bank. n.d. "Growth Will Slow but Remain Robust." World Bank. http://go.worldbank.org/CVKS5Q22G0 (accessed December 13, 2011).

———. 2001. *World Development Report 2000/2001: Attacking Poverty*. Oxford: Oxford University Press.

World Commission on Environment and Development. 1987. *Our Common Future*. Oxford: Oxford University Press.
World Council for Renewable Energy. 2005. "Renewable Energy and the City." World Renewable Energy Policy and Strategy Forum. Bonn, Germany. June 13–15.
World Council of Churches. 2005. "World Council of Churches Statement to the High Level Segment of the UN Climate Change Conference (COP11 and COP/MOP1)." Montreal, Canada. December 9.
World Tourism Organization and United Nations Environment Programme. 2008. *Climate Change and Tourism: Responding to Global Challenges*. Madrid: World Tourism Organization.
Wright, Erik Olin. 2010. *Envisioning Real Utopias*. London: Verso.
Wright, Steve. 2005. "Preparing for Mass Refugee Flows: The Corporate Military Sector." In *Clearing the Pathwys to Survival—The State, Ourselves, and Climate Change*, edited by David Cromwell and Mark Levene. London: Pluto.
Yamin, Farhana, and Johanna DePledge. 2004. *The International Climate Change Regime*. Cambridge: Cambridge University Press.
York, Richard, Brett Clark, and John Bellamy Foster. 2008. "Capitalism in Wonderland." *Monthly Review* 61, no. 1 (May): 1–18.
York, Richard, Eugene A. Rosa, and Thomas Dietz. 2009. "A Tale of Contrasting Trends: Three Measures of the Ecological Footprint in China, India, Japan, and the United States, 1961–2003." *Journal of World-Systems Research* 15, no. 2:134–46.
Zalik, Anna. 2008. "Liquefied Natural Gas and Fossil Capitalism." *Monthly Review* (November): 41–53.
Zhang, Zhihong. 2003. "The Forces behind China's Climate Change Policy: Interests, Sovereignty, and Prestige." In *Global Warming and East Asia: The Domestic and International Politics of Climate Change*, edited by Paul F. Harris, 66–85. London: Routledge.

Index